# PEER-LED TEAM LEARNING

# A Guidebook

## THE WORKSHOP PROJECT

*SPONSORED BY THE NATIONAL SCIENCE FOUNDATION*

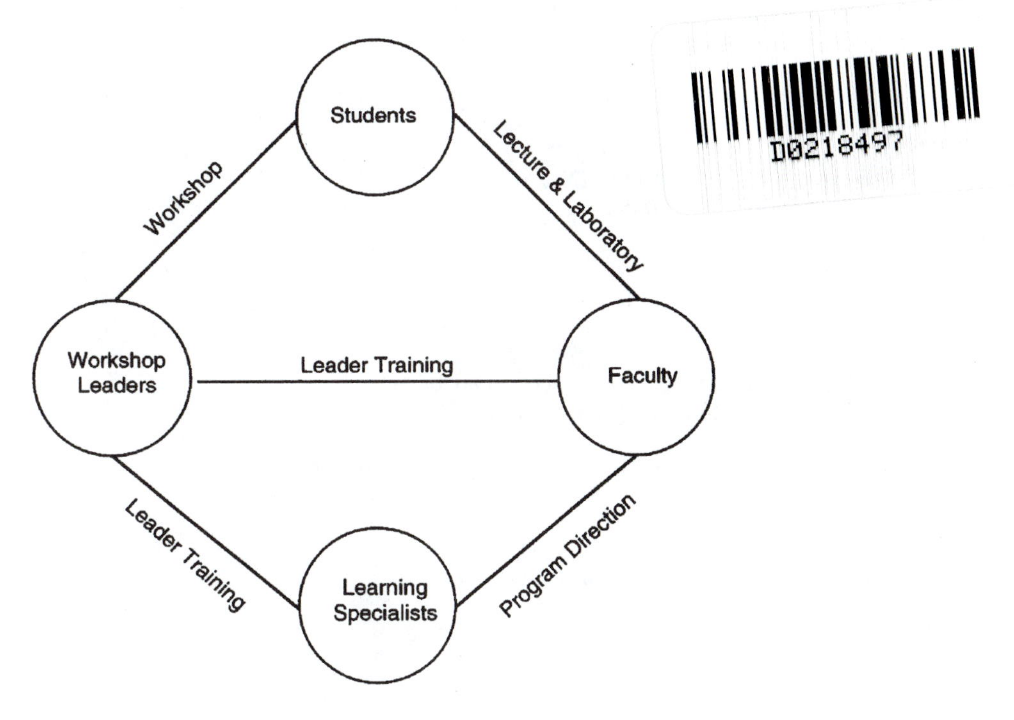



**DAVID K. GOSSER**
**MARK S. CRACOLICE**
**J. A. KAMPMEIER • VICKI ROTH**
**VICTOR S. STROZAK**
**PRATIBAH VARMA-NELSON**



***PRENTICE HALL SERIES IN EDUCATIONAL INNOVATION***

PRENTICE HALL, Upper Saddle River, NJ 07458

Executive Editor: ***John Challice***
Project Manager: ***Kristen Kaiser***
Editorial Assistant: ***Gillian Buonanno***
Special Projects Manager: ***Barbara A. Murray***
Production Editor: ***Benjamin St. Jacques***
Manufacturing Manager: ***Trudy Pisciotti***
Manufacturing Buyer: ***Michael Bell***
Supplement Cover Manager: ***Jayne Conte***
Supplement Cover Designer: ***Maureen Eide***


Upper Saddle River, NJ 07458



Printed in the United States of America

10 9 8 7 6 5 4 3 2 1

ISBN 0-13-028805-5

Prentice-Hall International (UK) Limited, *London*
Prentice-Hall of Australia Pty. Limited, *Sydney*
Prentice-Hall Canada, Inc., *Toronto*
Prentice-Hall Hispanoamericana, S.A., *Mexico*
Prentice-Hall of India Private Limited, *New Delhi*
Prentice-Hall (Singapore) Pte. Ltd.
Prentice-Hall of Japan, Inc., *Tokyo*
Editora Prentice-Hall do Brasil, Ltda., *Rio de Janeiro*

# *Preface to the Peer-Led Team Learning Series*

The Workshop Chemistry Project was an exploration, development, and application of the concept of peer-led team learning in problem-solving Workshops in introductory chemistry courses. A pilot project was first supported by the National Science Foundation, Division of Undergraduate Education, in 1991. In 1995, the Workshop Chemistry Project was selected by NSF/DUE as one of five systemic initiatives to "change the way introductory chemistry is taught." In the period 1991-1998, the project grew from the initial explorations at the City College of New York to a national activity involving more than 50 faculty members at a diverse group of more than 30 colleges and universities. In 1998-1999, approximately 2500 students were guided in Workshop courses by 300 peer leaders per term. In Fall 1999, NSF chose the Workshop Project for a National Dissemination Grant to substantially broaden the chemistry participation and to extend the model to other SMET disciplines, including biology, physics and mathematics.

*Peer-Led Team Learning: A Guidebook* is the first of a series of five publications that report the work of the Project during the systemic initiative award (1995-1999). The purpose of these five books is to lower the energy barrier to new implementations of the model. The *Guidebook* is a comprehensive account that works back and forth from the conceptual and theoretical foundations of the model to reports of "best-practice" implementation and application. Three other books provide specific materials for use in Workshops: *General Chemistry*; *Organic Chemistry*; and *General, Organic and Biochemistry*. One book in the series, *On Becoming a Peer Leader*, provides materials for leader training.

The collaboration of students, faculty, and learning specialists is a central feature of the Workshop model. The project has been enriched by the talents and energy of many participants. Some of their names are found throughout these books; many others are not identified. In either case, we are most grateful to all those who have advanced the model by their keen insight and enthusiastic commitment.

We also acknowledge, with pleasure, the support of the National Science Foundation, NSF/DUE 9450627 and NSF/DUE 9455920. Our work on the second NSF award was skillfully guided by our National Visiting Committee, Michael Gaines, Chair; Joseph Casanova; Patricia Cuniff; David Evans; Eli Fromm; John Johnson; Bonnie Kaiser; Clark Landis; Kathleen Parson; Arlene Russell; Frank Sutman; Jeffrey Steinfeld; and Ronald Thornton; we value their advice and encouragement. The text of the *Guidebook* was repeatedly processed by Arlene Bristol, with exceptional skill and remarkable patience. Finally, we appreciate the vision and commitment of John Challice and Prentice Hall to make this work readily available to a large audience.

Books are written for you, the readers. We welcome your comments and insights. Please contact us at the indicated e-mail addresses.

The Editors, Fall 1999

David K. Gosser gosser@scisun.sci.ccny.cuny.edu
Mark S. Cracolice markc@selway.umt.edu
J. A. Kampmeier kamp@chem.rochester.edu
Vicki Roth vrth@mail.rochester.edu
Victor S. Strozak vstrozak@gc.cuny.edu
Pratibha Varma-Nelson varmanelson@sxu.edu

# *Table of Contents*

# *Chapter One*
# *The Peer-Led Team Learning Workshop Model*

**David K. Gosser, The City College of New York**

An effective, proven model for teaching undergraduate science is the Peer-Led Team Learning (PLTL) Workshop, first used in teaching chemistry (Gosser and Roth 1998; Gosser and et al. 1996; Woodward, Gosser and Weiner 1993). The PLTL Workshop provides an active learning experience for students, creates a leadership role for undergraduates, and engages faculty in a creative new dimension of teaching.

In weekly two-hour Workshop sessions, students work together on challenging problems and discover that learning science is an intensely human activity, replete with joy and laughter, and struggle and frustration. Simultaneously they learn to take ownership of the material by using the language and ideas of science in focused discussions with their classmates. The model uses peer leaders to facilitate the cooperative work of small groups of six to eight students. These *Workshop leaders* are students who have done well in the course previously and are trained for their leadership roles.

The PLTL model is powerful. For the students, it increases their enthusiasm for the study of science and increases their success in the course. For the Workshop leaders, the experience of working with faculty and guiding others through a difficult course is unforgettable and can have profound effects on the leaders' growth. For the faculty, the model opens new dimensions of teaching, free from the constraints of the lecture. For all, the model creates a new sense of community. The PLTL model is robust and has been successfully employed in a variety of institutions, including community colleges, liberal arts colleges, technical colleges, large state and city universities, and private research universities.

This introductory chapter describes

- the basis of the Workshop model in learning theory, leadership, and team performance;
- the *Critical Components* of successful PLTL Workshops; and
- the impact of PLTL Workshops on students and student leaders.

The other chapters in this *Guidebook* were written by students, faculty, learning specialists, and evaluation experts who have contributed to the development and understanding of the PLTL Workshop model. Each section provides guidance to those who decide to launch student-led Workshops in their classes.

### Lecture and the Student Experience

Traditional instruction relies primarily on the lecture: a method of presentation of concepts, models, content, and problem-solving methods by an expert to a group of listeners and note-takers. In addition to providing students an expert view of the subject, the lecturer plays other positive roles, such as communicating a clear set of course expectations, professing enthusiasm and commitment to the subject, and modeling ways of thinking and analyzing. These are important outcomes, and thus the lecture maintains a significant role in the Workshop model.

Despite its traditional standing, the lecture/recitation method has limitations that are related to the lack of student involvement. In a typical lecture setting, the chairs are pointed (often fastened to the floor) in the direction of the lecturer. Even for the most active listeners, communication is primarily one way - from lecturer to student. Recitations are meant to be more interactive, but all too often they devolve into problem-solving lectures. Neither lecture nor recitation is a place where students can balance the receptive modes of listening with active participation in problem solving and scientific discussion and debate.

**The Reflective Practitioner**

Donald Schon coined the phrase "reflective practitioner" to describe the special kind of problem solving used by professionals to deal with complex problems (Schon, 1983). Such problems require the individual to interact with the problem in a reflective way because the problem cannot be solved by applying ready-made formulas. For the scientist, this means incorporating unexpected results and planning new experiments. For the architect, it means looking at the consequences of a proposed design and redesigning accordingly. For the teacher, it means listening to the students to understand their points of view; it means using flexible structures that respond effectively to the students' differences. Schon observed that many traditional college curricula are out of balance, with an overemphasis on the codification of knowledge and an underemphasis on the actual practices and processes that lead to independence and mastery.

An alternative to the top-down, lecture-only model of teaching is one that builds collaboration and peer leadership into the course structure. Although formal teams are used extensively in the workplace, they are a new concept in higher education. The PLTL Workshop model embodies the spirit of the problem-solving teams that characterize many modern enterprises (Reich 1991; Ainsworth 1999).

> *... symbolic analysts work in teams. Learning to collaborate, communicate abstract concepts, and achieve a consensus are not usually emphasized within formal education, however.*
>
> *Robert Reich,* The Work of Nations

**The Wisdom of Teams**

What are the "teams" to which Reich refers? They are reflective practitioners working together to solve complex problems. The *Wisdom of Teams* (Katzenbach and Smith 1993) is an intriguing study of actual teams in a wide variety of settings. Often, the teams were formed to confront daunting performance challenges. Through intense collaborative work, the team members developed the personal and professional resources and understanding that was needed to solve the problems. Whether it was the struggle to change the way railroads do business or to transform a little-known college basketball team into a national competitor, these small teams of about a half dozen participants achieved results that could never have been imagined by the individual members of the team. The essence of these high performance teams, according to Katzenbach and Smith, is "... a small number of people who are committed to a common goal, a common working approach, and to one another's personal growth and success."

**Teams and Learning**

Cognitive science identifies language as a central feature of learning. The process of expression, debate, discussion, and consensus is an effective way for students to learn the language of science and to construct their own understanding. Learning teams are a natural way to get students to talk to one another. They are particularly useful because students have compelling reasons to work together; they are faced with a difficult course and challenging problems. The diversity of viewpoints works to the advantage of the learning teams. The interchange in the group is like a distributed intelligence network, providing many opportunities to utilize different learning styles and to offer multiple representations.

During the first years of college, students have the greatest need to connect with others and become part of a community of learners. Although mentoring relationships are recognized to be important ingredients of interest and success in science, they are largely absent in impersonal introductory lecture classes. Workshop teams that are led by successful students build an extensive network of "proximal" mentoring that includes students, student leaders, and faculty. Theory indicates that the most effective learning takes place when assistance is offered by someone above, but near, the level of the learner's development (Vygotsky 1980; Tharp and Gallimore 1988). It is intriguing to note that the Workshop leaders, who are one or two semesters ahead of their group members, are especially well suited to help the students learn. These ideas are discussed in greater detail in Chapter 7.

Workshop teams provide opportunities for students to engage in reflective problem solving, to take risks, to become comfortable with the possibility of making mistakes, to check their understanding with colleagues, and eventually to triumph in mastering the course content. These are the same reasons that teams are so successful in other settings. The team members become resources for one another; the team leader guides and mentors the team members, providing opportunities at every turn. The support provided by the Workshop leader and the Workshop members helps each student reach his or her potential.

**An Untapped Resource**

College campuses have all the necessary elements to create Workshop teams to help students learn and develop critical thinking skills. Faculty are seeking to create exciting and compelling experiences for their students, students are eager to do well and to participate, and learning specialists know how to help fashion new environments for learning. The PLTL Workshop model makes use of the untapped resource in the college and university - the students themselves. The students in any class are resources for one another, as team members who build understanding through a mutual interaction and commitment, and, given the opportunity and provided with the appropriate structure and guidance, many students are ready, willing, and able to take a leadership role in guiding other students to learn (see Chapter 2).

**The Team Leader**

The PLTL Workshop shares elements with other curricular efforts involving a collaborative learning environment (Treisman 1992) and with proven pedagogical tools such as pair-problem solving (Whimbey 1982). However, the use of undergraduate students to take the role of team leaders, *Workshop leaders*, is a fundamentally pioneering undertaking (Woodward, Gosser, and Weiner 1993).

Because the team members are a diverse group with strikingly different approaches and skills to solve problems, the team leader has a special responsibility to guide the team to its full potential. According to Katzenbach and Smith "Team leaders act to clarify purpose and goals, build commitment and self-confidence, strengthen the team's collective skills and approach, create opportunity for others .... Team leaders do not believe that they have all the answers - so they do not insist on providing them."

*As the potential team grows ... the leader's job changes markedly. His or her formal authority may go unchanged, but when, whether, and how to use it shifts. The key to the leader's evolving role lies in understanding what the team needs and does not need from the leader to help it perform. The team leader is the ultimate utility infielder ... he or she must be there only to deliver as needed.* *J. Katzenbach and D. Smith,* The Wisdom of Teams

**The Workshop Model of Peer-Led Team Learning**

In practice the PLTL model introduces *a unique curricular structure: a weekly peer-led Workshop*. Each Workshop has six to eight student members and a trained leader. The Workshop leader guides the students in collaborative problem solving, model building, and discussion and debate of scientific ideas. Several years of Workshop evaluations (see Chapter 6) have identified six key elements of successful Workshops.

**The *Critical Components* of the Peer-Led Team Learning Workshop Model**

- **The peer-led team learning Workshop sessions are integral to the course and are coordinated with other elements.**
- **The faculty teaching the courses are closely involved with the PLTL Workshops and with the peer leaders.**
- **The peer leaders are students who have successfully completed the course. They are well trained and closely supervised, with attention to knowledge of the Workshop problems, teaching/learning strategies, and leadership skills for small groups.**
- **The Workshop materials are challenging at an appropriate level and, integrated with the other course components, intended to encourage active learning and to work well in collaborative learning groups.**
- **The organizational arrangements, including the size of the group, space, time, noise level, and teaching resources promote learning.**
- **The institution, at the highest levels of administration and pedagogy, and at the departmental levels, encourages innovative teaching and provides sufficient logistical and financial support.**

These six *Critical Components* of successful Workshops are explained in more detail next.

***Integral to the course.*** *Creating the Workshop community.* The nature of the Workshops is defined by the connections among those involved - the students, student leaders, faculty, and learning specialists who operate as a team responsible for the realization of the *Critical Components* (see Figure 1). The PLTL Workshop model not only creates a new role for undergraduate students as leaders but also initiates a collaboration between science faculty and learning specialists. Learning specialists have specific training and expertise in areas closely related to the theory and methods of the PLTL Workshop. There is significant potential for productive cooperation between faculty and learning specialists (who may be located in a learning assistance center or a faculty development center). Although a learning center may be able to provide logistical support for Workshops, the more important function for the learning specialist is to collaborate to create training sessions for Workshop leaders and to support faculty development in learning theory and practice. In turn, the collaboration provides a mechanism for building close working relationships between learning centers and the curriculum. These connections lead to increased communication and creative solutions for educators and students.

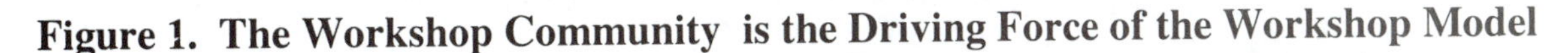

**Figure 1. The Workshop Community is the Driving Force of the Workshop Model**

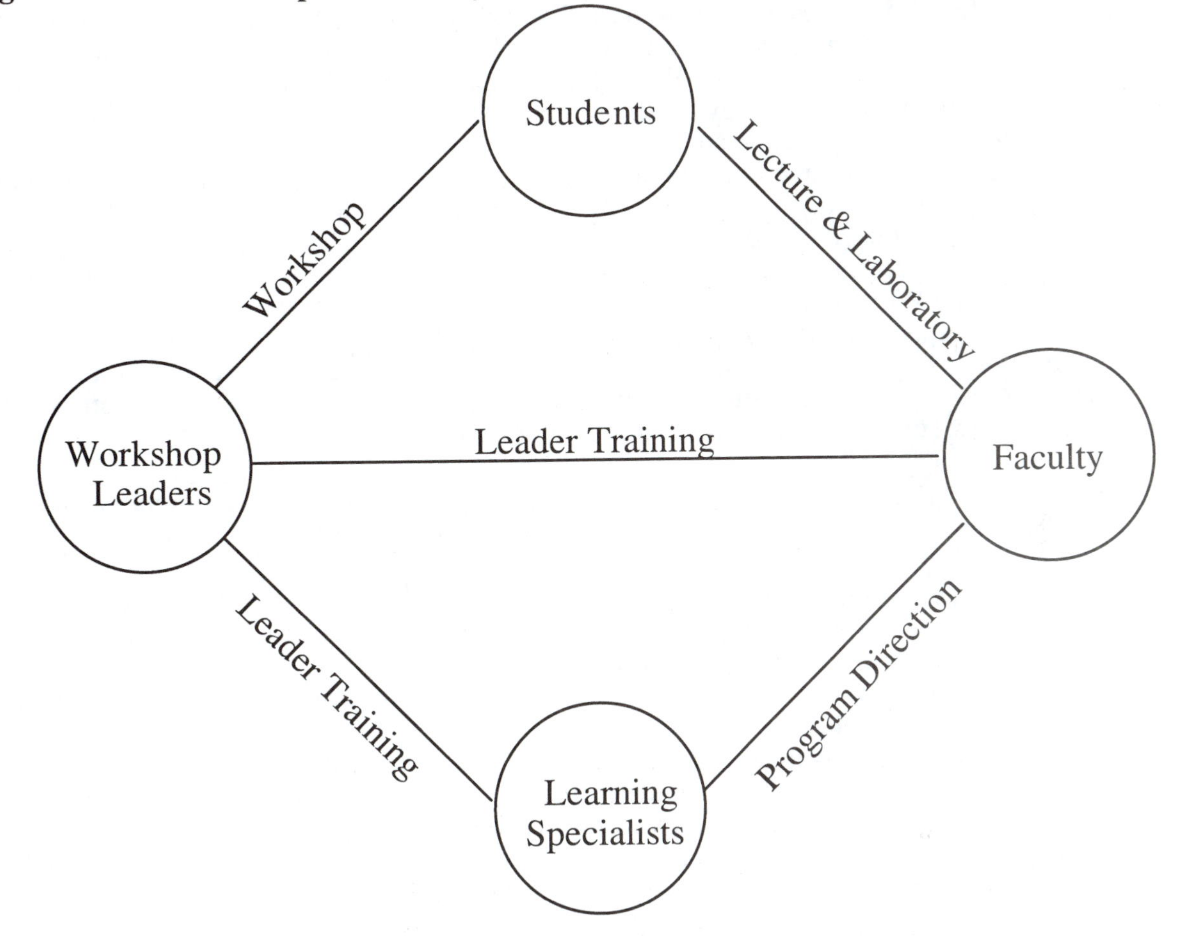


*The student-led Workshop is an integral part of the course.* At the beginning of the term, the students in a lecture class are organized into groups of 6 to 8 students, selected randomly. Each of these groups is assigned a Workshop leader; for example, a class of 100 would require 14 Workshop leaders. The leader is a student who has done well in the course previously and who has good people skills. The group meets on a weekly basis for the entire semester to engage in problem-solving sessions under the guidance of the Workshop leader. The Workshops are scheduled so that the necessary background has been discussed in lecture; the students are expected to have completed some preWorkshop assignments.

Keys to the success of the PLTL model are that the Workshop is built into the course structure and that the weekly Workshop sessions play a major role in the learning process. The Workshop is neither remedial nor optional. It is not a question and answer session. Rather, it is a carefully structured mechanism to help students build their understanding.

***Faculty***. In the PLTL model, the faculty become actively involved in creating and promoting the Workshop environment. They are responsible for ensuring that the Workshop materials are closely coordinated with course goals and with other course components. In their meetings with the peer leaders, they model ways to manage interpersonal dynamics within a team. They often guide the leaders through the upcoming Workshop, providing models for both content and leadership. They solicit feedback about the Workshops from the students and

Workshop leaders. The success of the Workshop inevitably has an impact on the teaching philosophy of the faculty. After seeing the benefits of an active learning environment, the lecturer often steps back to reexamine the role and the methods of the traditional lecture.

***Leaders.*** *The peer leader is central to the Workshop.* The peer leader has several roles. Each Workshop leader finds a particular style for a particular group; however, it is clear that the role is not that of lecturer or expert. The Workshop leader does not dispense answers. Assuming the role of guide or facilitator is the key to becoming an effective Workshop leader. The Workshop leader is there to guide the students to actively engage with the materials and with one another. This facilitation has several attendant methodologies, which are described in Appendix I. These include organizing round-robin problem solving, creating paired problem-solving groups, encouraging groups to compare results, offering timely assistance when a group is stuck, and providing encouragement and guidance about the course. A good leader knows when to help and when not to. The Workshop leader needs to set a tone for the discussion in which individual points of view are respected, criticism is constructive, and all members have equal opportunities to participate. Workshop leaders are successful students and often become mentors or role models to their groups. However, because they are close in age and only one or two semesters ahead of their fellow students, leaders remain nonauthoritarian.

> *This is the essence of the team leader's job: striking the right balance between guidance and giving up control.* — *J. Katzenbach and D. Smith,* The Wisdom of Teams

Although peer-led team learning is a small-group learning method and borrows from the large literature on cooperative learning and team methodologies, it is distinguished by the element of undergraduate peer-leadership. But why is a leader needed? Our view is that the peer leader liberates and empowers students to take responsibility for their own learning. Although the faculty are very much involved, they do not direct supervise the students during the Workshops. The instructor sets the content and materials and then trusts the team (of students and leader) to find its own way to deeper understanding. The highly structured and defined roles of team members described in the literature of cooperative learning are replaced by a responsive, flexible structure that has been found to be powerful in the context of the workplace; that is, a small group (a team) with a leader.

The leader makes the group coherent and effective by

- clarifying purpose and goals;
- ensuring full participation of the group members;
- building commitment and self-confidence;
- strengthening the team members' skills and approaches to problem solving; and
- creating special opportunities for team members.

*Learning the peer leader's role.* The Workshop leaders need to learn to do their jobs, and the faculty need to be closely involved in teaching the leaders. One part of this process involves preparation for the content of the Workshop. Although the Workshop leaders are expected to have reviewed the problems in the Workshop, it is important that they have the opportunity to discuss these problems with one another and with the faculty. In these meetings, the faculty member becomes guide and mentor to the student leaders. Although the context of the meeting is to discuss the Workshop problems, the faculty and leaders become more comfortable with one another; the leaders often turn to the faculty for advice about courses and career-related topics. In turn, the faculty turn to the leaders for feedback about the students, the course, and the Workshop sessions.

Being a Workshop leader requires more than content knowledge. The facilitator needs to know techniques for dealing with shy, dominant or unprepared students; teaching tools for working in small groups; ways to utilize different learning styles; and when and how much assistance to offer students. An explicit emphasis on leadership training can add immeasurably to the Workshop program. Trained leaders have more guidance, and thus their work is much more effective and rewarding. The training can be offered by the faculty or by the faculty in cooperation with a learning specialist. Journals that leaders keep can provide important feedback to the faculty and learning specialists and invite the leaders to reflect on their Workshop practices. Leaders learn to record and share their experiences with other leaders. The many dimensions of Workshop leader training are examined in Chapter 4 and in Appendix III.

There is a remarkable enthusiasm among students to become Workshop leaders; most find a great deal of satisfaction in the work they do. Indeed, the PLTL model has stimulated a larger commitment from some of the student leaders. They become involved in the project in a broader sense and contribute to the development of Workshop materials, work on Web pages, and participate in local and national meetings where they teach interested faculty by actually leading them through typical Workshops. These *super leaders* make tremendous contributions, injecting their own special brand of idealism and enthusiasm into the Workshop Project.

The care and attention that is given to the leaders is reflected in their relationship with the other members of the Workshop team. The following comments are from students who attended a joint faculty-learning specialist - Workshop leader training seminar at the City College of New York.

*Workshop leadership has a whole new meaning. It is something more than tutoring. It is fostering collaboration and helping students work together.*

*During the meeting, I had the opportunity to put myself in the shoes of a Workshop leader. However, I did so with the people who administer and coordinate the project. This really gave me a chance to feel like part of the driving force of the project. After the discussion on cooperative learning, I have come to realize how important and helpful group studying can be. By listening to other Workshop leaders, I learned about strategies that will be useful in future Workshops.*

***Materials.*** Each Workshop session is built around a set of problems and activities designed and structured by the faculty member to focus on the central ideas and to help the students attain the course goals. The PLTL model is quite flexible and can accommodate many different kinds of problems and materials. However, the Workshop environment is unique and invites some creative work and considerable rethinking of traditional problems. The problems must be designed to help the leader actively engage students with the material and with one another. Vygotsky (1980) emphasized that students learn when they are appropriately challenged. With the help of others (teacher, leader, fellow students), the student can reach new levels of accomplishment and understanding. The Workshop materials provide the challenge, and the Workshop team provides the help.

Problems based on the development of scientific ideas and thought allow students to explore some of the eclectic, intuitive reasoning that is often lost in the distilled wisdom of texts. These might include an exploration of the way Mendeleev created the periodic table or how van't Hoff figured out the tetrahedral geometry of carbon. Problems like these also reveal the epistemological bases of our disciplines.

In chemistry, problems that use molecular models are very effective because they tap into the kinesthetic and haptic dimensions of learning. The models are not mere learning tools but have played a significant role in scientific thought and discovery. In the process of exploring and manipulating the models, small groups of students can obtain an intuitive understanding for the constraints, interactions, and dynamics of systems that are difficult to intuit from equations, visual representations, or "untouchable" computer animations (Rouhi 1999).

Some PLTL Workshop materials have been specifically structured to enhance group interaction. Several examples are described next. More information on these methods and others, using Chemistry as the content area, can be found in Appendix I.

- A successful and widely used tactic is the round robin. The problem to be solved is broken down into a series of questions, and each member of the group is assigned one part. The method works quite well because each student has an equal voice in the discussion and because a successful solution requires that all students listen to one another.
- Concept maps (Novak and Gowin 1984; Herron 1996) that are created by small-group discussion are useful for reviewing topics such as chemical bonding.
- Workshop leaders report that "pair problem solving" (Whimbey, 1982) is a particularly useful pedagogical tool for getting students engaged. They often pair a stronger student with a weaker student, or a shy student with an outgoing student.
- Problems that involve concrete simulations and illustrate the nature of "play" in science can be utilized to provide insight into many areas of biology, physics, chemistry, and mathematics (Eigen and Winkler 1993).
- Finally, organizing the Workshops into teams that compete with each other in a *Jeopardy*-like tournament is an enjoyable and effective format to review factual material for exams (American Chemical Society, 1977).

*Integrating leader training, materials and methods.* At the beginning of each term, there is one key session at which the faculty meet with the Workshop leaders to prepare for the Workshops. The *Critical Components* of faculty involvement, training of leaders, examination of materials, and delineation of methods integral to the course are brought together. This is an important experience for faculty and peer leaders that ensures that they have a solid grasp of both content and pedagogical approaches. Creating materials that take full advantage of the PLTL approach and training the peer leaders to implement these materials are exciting tasks for the faculty. The instructor often leads a working session, using the materials that the peer leaders will use in the upcoming Workshop. This helps clarify the peer-leaders' understanding and illustrates models for leading the Workshop. The meeting also provides an important run-through to field-test the methods and the materials. The leaders' perspectives often reveal ambiguities in the materials, so they can be modified before they are used. In addition, new ideas for collaborative learning can be explored and tested in advance.

The following general hints are designed to help newcomers design and implement appropriate Workshop materials. More information is provided in Chapter 3. With the guidance of a trained leader, the group can quickly move, without trouble, into many different kinds of materials and modes of collaboration.

- Avoid lengthy explanations or complex instructions. The students have attended lecture, read the text, and worked some problems. The Workshop should take advantage of their prior knowledge and build on it.
- Do not start from scratch. Use your old problem sets and exams and modify their structure, content and purpose to fit the Workshop model.
- Aim for diverse responses to different learning styles by having students explain and compare results.

- Aim for quality, not quantity, in the PLTL Workshop. Doing a few representative problems well and allowing time for reflection and discussion is better than rushing through many examples. The latter will lead to the use of traditional didactic methods.
- Structure the problems and the training of leaders to minimize lecture presentation. Do not introduce new content in the Workshops or the leaders will end up lecturing.
- Make the methodology explicit to the leaders by example in the training sessions. The success of the training session is evidenced by the enthusiasm and participation of the leaders. If you lecture to your leaders, what do you think that they will do in the Workshop?
- Do not make answer keys! As tempting as this may be, it kills the Workshops.

***Organizational arrangements.*** Space and time are the major issues. The PLTL Workshop sessions require a space conducive to small-group discussion and work; a lecture hall will not do. A room with a table, chairs, and a chalkboard is ideal. Workshops must be clearly scheduled in advance and must meet on a regular basis. Campuses beginning PLTL Workshops may have to work on scheduling arrangements, to move through the layers of departmental and college approval. Once the success of the Workshops is apparent, this will be easier.

***Support.*** It is important for the administration to understand that the goals of the PLTL Workshop model coincide with the larger goals of the college and the department. The administration can provide support by recognizing and rewarding innovative and effective teaching. An administration that values PLTL Workshops will provide logistical and financial support as well. This support will ensure the institutionalization of peer-led learning teams. These issues are discussed at greater length in Chapter 5.

## Evolutionary Phases of a Workshop Team

While each group has its own individual style and way of doing business, the groups usually go through several phases during the term. Readings from the Workshop leaders' reflective journals tell us how the team members gradually recognize the challenges of learning science, learn the benefits of working together, and build a sense of personal commitment to one another.

*Phase 1: Trepidation on stepping into a new role.* The leaders often express some anxiety as they start to lead a new group.

> *My first day! At first, the butterflies in my stomach were all I could concentrate on, so I took attendance and said a few words about myself and organic chemistry so that the students would see me as part of the group and not some unapproachable Orgo genius.*

> *I definitely have to say that my first day as a Workshop leader taught me a lot about being an effective leader. There is more preparation than I had anticipated, and more patience than I had thought. How could I get them to see that I was not there to dictate answers but to act as their peer mentor?*

*Phase 2: Initial support for students.* The first exams often confront students with the challenge of learning chemistry, and leaders express their concern and support for the students.

*Since most of my students did not do well on the first exam, they were afraid of the second. I told them what happened when I took general chemistry a year ago. In my group, I got the lowest grade. Encouraged by my Workshop leader, who told me I could be a leader too, I started to study harder. So I did well in the course and now I am trying to help other students as I was helped.*

*Phase 3: The group starts to come together.* The leader, who initially has to take a very visible and prominent role in the discussion, finds that the students gradually become confident in group discussion and problem solving. They also take pleasure in noting the positive results.

*I was shocked to see my most shy student taking a very active role.*

*I saw something remarkable today that made me feel really proud. I was delayed and arrived late for the Workshop. When I got there, the students were not waiting for me, nor had they taken up their bags and left. Instead, I saw a student at the board, leading the group in question number one. When I arrived, I had only to say "carry on." This means that the group has achieved a certain level of independence: they can function as a unit on their own.*

*My group was not exceptional academically; however, they were superb at group dynamics. At the beginning of the semester, I found it necessary to subdivide the group into smaller cooperative learning sets. I also had to make my presence known in the workings of the group. Some of the men had to be needled to participate, while others had to be literally shut up. By the middle of the semester the group was working well without my direct guidance.*

*Phase 4: Validation of the team leader's effectiveness.* The outcomes and evaluation of the course methods are evidenced by the students' test results.

*I learned their test grades today and was flabbergasted by the results. One of my students who chronically does badly on the quizzes got a 91% on the test. Another, who never shows up for Workshop, got a 56%. I guess he doesn't understand the importance of Workshops.*

## The Peer-Led Team Learning Workshop Project

The Workshop Project is a coalition of faculty, students, and learning specialists committed to the PLTL model of teaching science. The project has been supported by the National Science Foundation since 1991. In 1998-1999, more than 50 faculty and 300 leaders conducted PLTL Workshop courses at more than 30 colleges and universities for 2500 students each semester. The guiding members of the project are committed to developing the peer-led team learning model by offering assistance to others who want to start their own Workshop courses. In 1999 the National Science Foundation awarded the PLTL Workshop Project a National Dissemination Grant to support the spread of the model to other chemistry programs and to other disciplines, such as biology, physics, and mathematics.

This *Guidebook* provides a comprehensive introduction to the Workshop model. The peer-led team-learning model has much to offer: increased student enthusiasm and performance, an exciting new leadership role for students, a new sense of community in science and other departments, and a new dimension in teaching for faculty and learning specialists; however, it is often very difficult to introduce change, even the modest change proposed here. The information provided here is not only a *how-to* guidebook but also provides ammunition needed to convince colleagues and administrators that the PLTL Workshop is a cost-effective solution for increasing student performance.

## Future Opportunities

There are several exciting challenges before us. We invite you to join us in the exploration and development of these new ideas and opportunities in peer-led team learning.

- Encourage new faculty, students, and learning specialists to implement Workshop courses and to participate in the Workshop Project activities,
- Expand the PLTL model to include biology, physics, mathematics, and earth and atmospheric sciences,
- Increase connections with teacher preparation programs and with high schools;
- Learn from other science initiatives that promote the use of information technology, real-world questions, discovery learning, and guided inquiry (see World Wide Web pages: Curricular Reform Projects),
- Continue research on the development and evaluation of the PLTL model.

## The Peer-Led Team Learning Workshop Project Web Page

The PLTL Workshop Web page is designed to provide up-to-date information on project Workshops and presentations, and available materials, and to encourage the development of an interactive database of Workshop implementations and evaluation.

**http://www.sci.ccny.cuny.edu/~chemwksp**

## WWW Pages: Curricular Reform Projects

Bioquest
http://www.beloit.edu/bquest

ChemLinks
http://chemlinks.beloit.edu

Long Island Consortium for Connected Learning
http://www.licl.org

MADCP, Middle Atlantic Discovery Chemistry Project
http://madcp.fandm.edu

Mazur Physics Education
http://mazur-www.harvard.edu/education/EducationMenu.html

ModularChem Consortium ($MC^2$)
http://mc2.cchem.berkeley.edu

Molecular Science Initiative
http://www.pslc.ucla.edu/MolSci.html

New Traditions
http://genchem.wisc.edu/newtrad/index.html

Problem-Based Learning
http://www.siumed.edu/pblc/index.html

Process Workshops for General Chemistry
http:\\www.chem.sunysb.edu/hanson-foc/index.html

Project Kaleidoscope (PKAL)
http://www.pkal.org

**References**


Ainsworth, S. J. (1999). Teamwork. *Chemistry and Engineering News 77* Nov. 15, 54-59.

American Chemical Society Satellite Television Seminar (1997). *Teaching Chemistry: Undergraduate Chemistry Curriculum Reform.* Available from Department of Continuing Education, American Chemical Society, 1155 Sixteenth Street, N.W., Washington, D.C. 20036.

Eigen, M. and R. Winkler (1993). *Laws of the Game: How the Principles of Nature Govern Chance.* Princeton, N.J.: Princeton University Press.

Goroff, N. (1998). A Report on Workshop Chemistry. The Chemical Educator. *The Chemical Educator 3,*1, [on-line serial]. Available at URL:http://journals.springer-ny.com.chedr/

Gosser, D. and V. Roth (1998). The Workshop Chemistry Project: Peer-led Team Learning. *Journal of Chemical Education 75:* 135-137.

Gosser, D., V. Roth, L. Gafney, J. Kampmeier, V. Strozak, P. Varma-Nelson, S. Radel, and M. Weiner (1996). Workshop Chemistry: Overcoming the Barriers to Student Success. *The Chemical Educator 1,* 1 [on-line serial]. Available at URL:http://journals.springer-ny.com.chedr/

Herron, J. D. (1996). *The Chemistry Classroom: Formulas for Successful Teaching.* Washington, D.C.: American Chemical Society.

Katzenbach, J. R. and D. K. Smith (1993). *The Wisdom of Teams.* New York: HarperBusiness.

Novak, J. D. and D. B. Gowin (1984). *Learning How to Learn.* Cambridge: Cambridge University Press.

Rouhi, A. M. (1999). Tetrahedral Carbon Redux. *Chemical and Engineering News*, *77* Sept. 9: 28-32.

Schon, D. (1983). *The Reflective Practitioner: How Professionals Think in Action.* New York: Basic.

Tharp, R. G. and R. Gallimore (1988). *Rousing Minds to Life: Teaching, Learning, Schooling in Social Context.* New York, Cambridge: Cambridge University Press.

Treisman, U. (1992). Studying Students Studying Calculus: A Look at the Lives of Minority Mathematics Students in College. *College Mathematics Journal 23*: 362-372.

Whimbey, A. and J. Lochhead (1982). *Problem-Solving and Comprehension.* Hillsdale, N.J.: Erlbaum Associates.

Woodward, A., D. Gosser, and M. Weiner (1993). Problem-Solving Workshops in General Chemistry. *Journal of Chemical Education 70*: 651-665.

Vygotsky, S. (1980). *Mind in Society.* Cambridge: Harvard University Press.

# Chapter Two
# The Students' Perspectives

At the heart of the PLTL model is teamwork, not only during the sessions themselves but also in the way we think of the instructional staff. To succeed, this model needs to have a team behind it, and crucial to that instructional team are the students who guide the groups. We do not refer to these students as *helpers*, as in the common term *teaching assistants*. Instead, we call them *peer leaders*. They are the leaders of the Workshop teams; they are the distinguishing feature of the PLTL model.

In the course of the project, we have listened carefully to the experiences and ideas of our students and our peer leaders. They have taught us about how students react and think, about the strengths and weaknesses of the materials we have designed for the Workshops, about our lectures, about the inner workings of a typical Workshop, and about their own personal development. In addition to their insights into the process of learning, our leaders have revealed a remarkable generosity of spirit, good will, and support for their fellow students. They have given us a view of our students at their best. For these reasons, we present the leaders' voices here, speaking from their dual perspectives as students and as leaders. We also include a few student voices. You will hear from students from different backgrounds and different types of institutions. The following reports come from the front lines.

**Gaya Amarasinghe, City College of New York**
*General Chemistry*

My association with the Workshop Chemistry Project began as a student in the freshman chemistry course at CCNY. I would like to outline some of the important outcomes of the Workshop Project as seen through my own experiences as an undergraduate student at City College and more recently as a graduate student teaching assistant in the Chemistry and Biochemistry Department at the University of Maryland at Baltimore County.

Being a part of the Workshop chemistry groups as a student, I was able to take advantage of the obvious: gaining a better understanding of the material. For me adjusting to the Workshop group environment was somewhat difficult, because I had just arrived from Sri Lanka, where studying alone was the norm. The Workshops became my first formal introduction to working in groups. In addition to learning chemistry, I was able to develop a number of important skills such as communication, team work, and leadership. In the realm of "hard subjects;" chemistry is likely to be a close second to physics in the courses-to-be-avoided category. From my own experience as a Workshop leader for five semesters, and a leader coordinator for two semesters, Workshops do not in any way make chemistry a "baby subject." Instead they create an atmosphere where students can interact with one another and use one another as resources. At the Workshops I was able to ask questions (and answer some) because I felt that everyone in the group was equal. To some, this process may seem like "the blind leading the blind." In my opinion, this is where the Workshop leaders are critical. One of the primary responsibilities of the Workshop leader is to guide the group. During my period as a leader, I saw my groups become progressively more independent. This was a result of my experience and ability to guide the group better. One semester I had as many as three students join the program as Workshop leaders.

Although I am no longer associated directly with the Workshop Chemistry Project, I see its influence in my work. Most recently, I was able to use my Workshop leader experience in my TA role in organic chemistry. The first week of the courses, I gave my students the choice of having discussion-oriented introduction to each lab or a lecture introduction. In the discussion-oriented method I used the same type of questioning used in Workshops to guide students to come up with the procedure and a flowchart to the lab instead of giving them a "things to do list." By the end of the semester, the students came to lab prepared and knew what to do. It not only made my work a lot easier but the students learned a little bit more in the end.

**Jewel Daniel, City College of New York**
*General Chemistry*

Since entering the City College of New York three years ago, I have been involved with the Workshop Chemistry Project on several different levels. It has been an eye opening experience. As a student taking the course, as a peer leader directing Workshop groups, and as a coordinator disseminating the good news, I have benefited tremendously and learned a great deal. Now as a high school biology teacher, I can and do apply some of these techniques that I learned from the chemistry Workshops.

I can recall my first semester at City College. I was a transfer student from the University of the Virgin Islands, a school with fewer than 5000 students. Just looking at the size of the Chemistry class overwhelmed me. But on that first Friday, we were placed in small Workshop groups. For me that provided the setting I needed for understanding the material. No longer was I sitting amid a hundred other students listening to a professor spout the words of knowledge; I was now a part of my own learning. The cooperative setting gave us the opportunity to do hands-on work. I had seen many of the concepts we covered in the course before, but I had never understood them as I did after being in the chemistry Workshop.

By the next semester, I was a peer leader with my very own group of students. I learned the art of facilitation. This meant that I had to prepare the material beforehand by working the problems, and I had to learn to needle the students to participate and contribute constructively to the discussion. It was from leading the group that I realized that one has not learned until one has taught. Not only did I understand the material better, but I was fast becoming an effective communicator.

Before assuming the role of Workshop leader, I thought the duties were centered around working chemistry problems with the group. When I became a leader, I realized the role was much more expanded and developed my leadership skills. Leaders constantly had to evaluate the Workshop materials. After all, we were the soldiers on the front line, constantly in contact with the students using the materials. Troubleshooting was commonplace. I recall on several occasions having to abandon the suggested method of problem solving and create my own. By the end of my time as a Workshop leader, I was able to think critically and make wise decisions quickly. I was also a much better leader than I had been initially.

At City College of New York, all Workshop leaders were required to take a course in leadership training. At first, many leaders considered it just a time-consuming burden. However, the training course gave us many of the skills that became essential as our roles expanded. We learned different ways of assessing student success. Integral to the course was the reflective journal that we had to keep. For the first time in my life I found myself reflecting on my actions. I was able to see my own growth and personal development. I was able to see the effect that my action, my personality, and my qualities had on the students in my group and on my fellow leaders. In addition, the course provided a support base for the new leaders who were uncertain about their role.

Being a student taking part in the Workshop Chemistry Project and being a peer leader convinced me of the effectiveness of the project. After a few semesters of being a Workshop leader, I became the coordinator of the leaders. In this role I had the opportunity to present my experiences with the Workshop Chemistry Project at different conferences. It was during this time that I became aware of the many similar curricula reform projects that were being implemented in different schools. My impression of college faculty was greatly altered as I began to get a sense of a greater community among educators. Speaking to faculty at the many conferences improved my communication and social skills greatly. On a local level, the process of recruiting and managing leaders helped me to understand what employers look for in prospective employees. I was able to reflect on my involvement in the project and assess its effects. I was able to grow in responsibility.

I graduated from City College in June 1997 and found the skills that I obtained as part of the Workshop Chemistry Project integral to my job search. I am presently a high school biology teacher. In my work, I find myself constantly drawing on many of the techniques of peer-led team learning that I garnered during my Workshop chemistry years.

Looking back at my college life and my involvement in the Workshop Chemistry Project, my favorite saying comes to mind, "The journey has not ended, the battle has not been won, till you look back into the mirror and see how far you've come." I am now looking back into that mirror. I see improved social and communication skills. I see improved understanding and retention of chemistry concepts. I see exposure to pedagogy on a grand scale. I see effective teaching tools. I see personal development and a love for education. Most of all, I see a very effective method of teaching in the Workshop Chemistry Project.

**Andrew Johanek, St. Xavier University, Chicago**

*This was written as Andrew was about to begin his Workshop leader assignment in organic chemistry.*

The Workshop provides an opportunity for students to get together as a group and use their collective knowledge to solve problems. Workshops are important because they remove the fear of giving the wrong answer. In my Workshop group there were two or three students who hardly spoke in class, but these same students regularly participated in Workshop discussions. Students like these build confidence over the year through the Workshop discussions. I think it was the ability of the students to discuss and explain things in their own way that was so important. Many times a fellow student brought a unique perspective to the material that helped me and others understand it.

When I first began organic, I was indifferent to the idea of Workshops. I did hear some students complain that it should be solely the teachers' responsibility to help them learn. They thought they were paying a lot of money to teach themselves. I was a little different; I was a victim of premedical syndrome. I wanted to learn only what I needed to know to get an A and do well on the MCAT. I had a rough time at first because I wanted to impress my professor almost as much as I wanted to understand the material. I had little concern for my fellow students.

As I became a bit more comfortable with the chemistry and with the Workshop model it became easier for me to sit back and listen to other people answer questions. I even began trying to think of questions I could ask the other students to help guide their thoughts. Asking the right questions helped me and the other students understand the material better.

As a leader I hope to be in tune to how my group is dealing with the material. I will try to promote the formation of the same group bond that we had in my Workshop group. I will not let any one person dominate the Workshop, and I will do my best to make sure that they understand why this is so important. My goal is to be as good as my Workshop leader was.

**William Mills, University of Rochester**
*Organic Chemistry I*

When I first learned that my sophomore organic chemistry class at the University of Rochester would be supplemented by a "Workshop," I envisioned power tools, a cluttered workbench, and sawdust. I was unsure how I was to learn chemistry in this seemingly garage-like setting. After attending my first few Workshop sessions as a student. I began to realize that my preconception of the Workshop program was not too far from the truth. Big, fat packets of hieroglyphic-laden pages, sets of plastic balls and sticks mixed on a tool bench-like surface formed by several pushed-together desks, stacks of used and unused scratch paper and orange pencils, and the low buzz of students talking - sometimes to other students, and often to themselves - contributed to a scene not unlike a brainstorming session in a real Workshop at an industrial company. By the end of the second session I began to feel very comfortable with learning chemistry in such a setting, and so did my classmates. The collegial atmosphere, the team approach to problem solving (often by dissection and distribution of small parts of a complex problem to individuals in a group), and the gratification of sifting through the seemingly incomprehensible chemical jargon to solve a practical word problem ingrained organic chemistry firmly in our minds. People in my Workshop seemed to become comfortable with approaching the most ominous problems by term's end. Having the ability to talk over imposing material in an informal setting helped me gain confidence in my problem-solving ability and fully prepared me for examinations.

The following year, the tool belt and hammer were passed to me. As the year began I was truly excited about being a Workshop leader. I was eager yet not completely sure what my exact role was to be. Through the required leader support class, and mostly by experience, I quickly came to understand how I could best help the students. I would be a guide, not a teaching assistant. I would listen, not lecture. At times, biting my lip was difficult. As the term progressed I found that listening to the students and dropping occasional hints when a group seemed stuck was a very effective strategy. However, there were a few instances in which none of the three or four people in a group seemed to be headed in the right direction. Surprisingly, ideas seemed to flow, and a chain reaction often ensued, leading to the correct solution with no direct help from me. I witnessed students with little chemical confidence at the beginning of the term become adept problem solvers. The students in my group seemed to enjoy their time in Workshop, and I believe they put forth greater effort due to the Workshop approach.

I learned more chemistry during my time as a Workshop leader than I would have thought possible. Perhaps I learned even more than I would have as a traditional teaching assistant because of the need to pay such close attention to what the students were saying rather than spouting my own knowledge. It takes a very deep understanding of the material and personal restraint to play the mental game of chess involved in being an effective Workshop facilitator. To prove how much knowledge I gained from the experience, I took the comprehensive review for the final exam as a student and then again as a leader. As a leader, I was able to answer every question correctly in a short time. This is something I definitely could not have done the previous year.

Being a Workshop leader benefited me in many ways in addition to making me more proficient at organic chemistry. It helped my people skills, my confidence, and my patience. I understand the role of a "teacher" in a new way after years of being a student. The understanding I gained has given me a new appreciation for those teachers who take time to listen to their students and who realize that often the best way for students to learn is not by didactic presentation but by allowing students to make their own intellectual mess and to learn how to clean it up effectively and efficiently. There is no substitute for a hands- (or brains-) on experience of trial and error.

**Carol Munch, St. Xavier University, Chicago**
*Organic Chemistry*

I recently completed a full year of organic chemistry at St. Xavier University. I had the unique opportunity of taking a full year of general chemistry some twenty-five years earlier at a much larger institution, the University of Illinois in Champaign, Illinois.

I eventually went on to complete an accounting degree at DePaul University and spent the next twenty years as an international tax practitioner both in public and private accounting firms and in large corporations. I returned to St. Xavier University recently to complete the requirements necessary to teach in the field of science.

In returning to school I noticed that the way chemistry is taught had changed dramatically over the last twenty-five years. I was extremely lucky to participate in a very different way of learning organic chemistry. We met on Mondays and Wednesdays for the usual one-hour lecture. On Friday, the format was changed: we met in a small peer-led Workshop group for two hours.

My organic chemistry professor will acknowledge that I was not a strong proponent of peer-run Workshops. I was a firm believer that chemistry, especially organic chemistry, was a subject best taught through lecture and repeated independent study. I was adamant that lectures, along with the numerous resources available on the Internet, were a better learning tool than any peer-led Workshop. What could a student who had just completed a year of organic chemistry teach me in a Workshop? After a full year of Workshop participation, I reversed my opinion about peer-led Workshops.

Someday, I hope to teach science. Why one method works over another is of special interest to me. For that reason I have tried to summarize why I believe the addition of peer-led Workshops to certain sciences, especially organic chemistry, is an effective teaching tool.

In our small Workshop setting, I found that students who did not participate in lecture or classroom discussion did participate in a Workshop. After a few sessions, students began to recognize that they alone were responsible for the material. No one was "spoon feeding" them. During the first few Workshops it became clear that most of the students were operating at the same level of uncertainty. I asked questions because I needed to know the material. I answered questions to obtain confirmation that I was heading in the right direction. Problem solving was addressed in an organized fashion in the Workshops; we learned a systematic approach to problem solving. Organic chemistry is cumulative. By watching our participation in the Workshop, our peer leaders recognized areas in which individual students needed additional attention. Without this intervention it was usually would be exam time before a problem was noticed.

I had a different peer leader each semester. Both peer leaders were well aware of the errors we made as students, having made the same errors and experienced the same problems just the year before. Both peer leaders were fair and nonjudgmental. Although they maintained an active profile as the peer leaders, they still made us do the problems. Each acted more as a facilitator than as a leader.

We used the Workshop as a communication forum. Science majors, myself included, are not known for their interpersonal skills. The Workshop gave us a chance outside of the larger lecture hall setting to exchange ideas on reaction mechanisms, on synthesis problems, and even on lab experiments results. I know that participating in the Workshop better prepared us to exchange ideas without criticism.

Finally, I strongly believe that peer-led Workshops prepare the student for the years after college. I spent a number of my working years as a manager in three of the largest public accounting firms. During those years I was involved in hiring new accounting graduates. Like the graduate and medical schools, the better accounting firms want to hire only the accounting students with the highest GPAs. Unfortunately, there was no correlation between a student's GPA and his or her ability to work well in a business environment. In today's corporate environment, success is measured not only by personal performance on the job but, more importantly, by what the individual contributes to the team and to the corporation as a whole. Teaching peers to work as a team in Workshop prepares them to be successful in group situations after college. Although I am certain we will forget much of our organic chemistry, I think we will not forget how to go about learning new concepts or how to work together.

**Elna Nagasako, University of Rochester**

I am currently a project coordinator in the Department of Anesthesiology working on a study of pain sensitivity and affective deficits in schizophrenia. I am also finishing up my Ph.D. in optics in the area of spatial soliton propagation. I have been volunteering at a local psychiatric hospital and am considering applying to medical school with the intent of working with the mentally ill. I took a year of organic chemistry in order to fulfill the admission requirement for medical school. I found the Workshop enjoyable and productive for a variety of reasons.

*The Workshops increased my interest in the subject material.* An opinion that I have heard expressed in discussions of science education is that any course change that makes students more enthusiastic about the material must also involve "dumbing down" the material. I thought it was great that the Workshops in organic chemistry defied that assumption. I felt that the Workshop problems were challenging and that the additional factor of group discussion kept me alert and motivated to explore the material.

*The Workshops encourage questioning and discussion.* One thing that has become clear to me after my time in graduate school is the importance of being able to communicate and defend scientific ideas, yet nothing in the format of standard science classes promotes this skill. I think one benefit of Workshops is that they are a gentle introduction to the sometimes combative arena of scientific discussion. Participants are encouraged to state their ideas and to question others in a nonthreatening environment. These discussions build communication skills and deepen the participants' knowledge of the material.

*The Workshops provide a social framework for the class.* Given my nontraditional background, I began the semester knowing no one else who was enrolled in organic chemistry. This was of concern to me, since I felt that without knowing other students it would be easy to go off track. Knowing other students provides an information "buffer," a way of ensuring that time is not wasted or confusion caused over a missed piece of information. The Workshops provided me a ready-made chemistry network that was invaluable and would have been difficult for me to arrange otherwise. I would imagine that this would be true for other nontraditional students as well as for those undergraduates with an interest in chemistry but whose social network does not include other science majors.

**Dawn Patitucci, St. Xavier University, Chicago**
*Principles of Organic and Biochemistry*
Organic Chemistry I and II

I had just completed a chemistry course (Principles of Organic and Biochemistry) for allied health students when Dr. Varma-Nelson contacted me and asked if I would be interested in working as a peer leader for a new project she was going to initiate in fall 1995. It turned out I could not be a peer leader due to a schedule conflict, so I became a tutor for two hours a week. In the following semester, I became a peer leader in a Workshop. What follows is a comparison of my experiences as a tutor with those as a peer leader.

While I was tutoring I was very rigid in my approach: if you want good grades, study. It was all very cut and dried. This attitude was the driving force behind my tutorials. The students would be seated, and I would stand up at the board and pretend that I was the teacher and "spoon-feed" them what they needed to know. I had a group of regulars who became rather dependent on me. They would tell me, "When I'm here with you, I understand, but as soon as I walk out of here, it'll be lost." I thought that meant I was doing a great job. It wasn't until the end of the semester, when most of them had to drop the class, that I realized that it wasn't as simple as I had thought.

The following semester was my first as a peer leader. Because there were not enough peer leaders available, I was assigned a group of twelve allied health students. Twelve students is a bit too many - the model calls for six to eight. If I wasn't careful, the group would almost take on the shape of a small class, with all of them facing me and talking to me instead of to one another. Of those twelve students assigned to me for that semester, four or five of them happened to be excellent students, and the rest were also academically strong. This was a good thing for me because I was still struggling with the impulse to play teacher. It was difficult for me to sit back and let the students figure things out for themselves without running to the board to solve problems for them. Soon I realized how capable this group was, and was able to let go.

I find this model to be beneficial because it encourages communication and teamwork among students. It also provides an escape from the monotony of lecture. The absence of the professor and the small group setting creates a comfortable environment where students can discuss the material in their own language. That, I believe, is an important part of the learning process. It is quite clear to me, as I see the students in my group adjust to the course, that what they will get out of the Workshops is directly proportional to the effort they put into the Workshop sessions. This is the case with just about anything in life, which makes the Workshop experience all the more valuable.

The benefits of being a peer leader are many. The knowledge I've gained has been invaluable to me in my other courses. My problem-solving and communication skills have improved, and I have developed closer ties with my peers. There have been times when I wasn't so sure I wanted to do this. Dr. Varma-Nelson convinced me not only to stick with it but to take the opportunity and run with it. These are all objectives of the Workshop model, and I'm amazed that these things happened to me because the only thing I wanted from St. Xavier University when I arrived was to get a degree and get out. This model really helped me to carve out a niche for myself at St. Xavier. In May 1997 I graduated, but I continue to work as a peer leader and Workshop coordinator for organic chemistry at St. Xavier University. When I graduated I was not ready to simply walk away from the project which had become such an integral part of my life.

# *Chapter Three*
# *Writing Workshop Materials*

**Victor S. Strozak, New York City Technical College**
**Donald K. Wedegaertner, University of the Pacific**

Peer-led team learning (PLTL) is a transforming idea that changes the relationships of student and faculty to the subject matter of the course and to each other. The goals are new levels of *individual* learning, understanding, and self-reliance. The structure of the PLTL model is *cooperative* problem solving under the guidance of a trained peer leader. The PLTL Workshop structure is unique; in turn, the model requires problems that are designed to fit the structure. We all know the traditional kinds of problems in our courses and textbooks. In the main, they are designed for students working alone. The PLTL model transforms our understanding of good problems; the transformation is a consequence of the structure of the PLTL Workshop. The structure also opens up new dimensions in the exploration of a given problem.

The importance of well-designed Workshop materials is identified as one of the *Critical Components* of the PLTL model. That analysis gives us a straightforward guide to the essential characteristics of good Workshop materials they should be:

- appropriately challenging;
- integrated with other course components and goals, including exams; and
- designed to engage the students with the material *and* with one another.

A quick check will show that most traditional problems fail to meet one or more of these criteria. Although short-answer drill problems have their place, they do not provide the challenge that leads to new conceptual understanding. The appropriate challenge lies beyond the individual but is within the reach of the group. The challenge does double duty by stretching the participants and requiring them to rely on one another.

Students have much to do, and they carefully guard their time and energy. In order to win their active engagement, the Workshop materials must be perceived as new and different from other problems, relevant to learning the subject and doing well in the course, and intimately connected to other parts of the course work. It also helps if the problems are fun to do and provide a sense of accomplishment in the end (e.g., a puzzle solved, a formula constructed, a map developed).

Finally, the PLTL model is grounded in the power of the team. The social structure of the Workshop provides the essential stimulation, support, and challenge that lead to new learning and developing. The engagement with others helps the individuals construct new personal understanding. The Workshop materials must facilitate this process by structuring activities that require the participants to interact.

Writing Workshop problems and materials has been one of the major activities of the members of the Workshop Project. The materials have been tested through several cycles of student use, often with different faculty and in different colleges and universities. The problems have been refined and revised as a result of field experience and with advice from students and leaders. The end result is a series of three books that provide complete, year-long collections of Workshops for typical courses in general chemistry, organic chemistry and allied health chemistry (general, organic, and biochemistry). These books are a central part of the Workshop Chemistry Project Series.

We learned the answer to two important practical lessons as we put the Workshop materials into practice. First, should Workshop problems be distributed in advance or simply be handed out in the Workshop? Second, should answers be provided to the leaders or to the students during the Workshop? We learned that it is best to hand out the problems ahead of time and to explain that the Workshop problems are not homework; rather, they are to be worked on in the Workshop, with other students. At the same time, we emphasize that students need to prepare for Workshops. The previous week's lectures, text, and problem assignments are the necessary first steps in preparation. Most good students will take a look at the upcoming Workshop as part of their preparation; they may even use the Workshop to direct some of their studying. The Workshop materials must be recognizable as subsequent steps in learning.

The second common question concerns answers. Students are eager to have the *answers*. Beginning leaders ask about answer keys. Faculty wonder how to respond. We made a conscious decision not to provide answer keys to Workshop problems, either to leaders or to the students. Our reasoning was that students have always had access to lots of answers, to both text problems and to exam questions. If it were sufficient to simply provide the "right" answers, none of us would be involved in the Workshop Project. The trap in solution manuals is that many students can recognize the answer when they see it. Students confuse *recognition* knowledge with understanding.

The Workshop Project is about learning how to construct answers, how to evaluate different answers, how to test for ambiguity, and how to test for completeness. In the end, it is about the process that allows students to assert, with confidence, "This is the best answer." It works best to grapple with these issues in a team setting because the process requires a discussion of a range of ideas; debate about the validity of the ideas, checking, testing and correcting ideas; and building consensus and confidence. Ultimately, students internalize these processes and practice them. Simply put, the value of the Workshop and the special role of leaders would be short-circuited by an answer key.

We also learned a lot about writing Workshop materials. We often started with a traditional problem or a favorite exam question. Invariably, these problems were shaped and modified to fit the special requirements of the Workshop structure for student participation and debate. Since we wanted the Workshop problems to focus on **big ideas**, the process forced us to think carefully about what we were trying to teach. Most problems evolved as we got feedback from students and leaders. To be sure, there is much to be done to get a good set of Workshop problems. Nevertheless, we found that the process sharpened our thinking about the content of our courses, the importance of different ideas, and the ways that our students learn. As a result, we found the process stimulating and rewarding. The following list organizes some characteristic elements of good Workshop problems:

1. *Focus on developing conceptual understanding.* Avoid drill.

2. *Explore only one or two big ideas in a connected, coherent fashion.*

3. *Encourage students with different abilities. A graded approach works best; start easy and work to hard*

4. *Foster group interactions.*
   Some activities meet objectives 3 and 4 simultaneously:
   - multistep problems that are approached in round-robin fashion
   - puzzle problems in which students list the *observations* and corresponding *deductions* one-by-one. Putting the deductions together to get a conclusion is the hardest part.

- problems in which a group of two or three students build molecular models, examine the models, and discuss the implications of the structures

5. *Recognize different learning styles.*
   Different types of activities that accomplish this include:
   - building molecular models
   - brainstorming (discussing) a problem
   - explaining to others
   - constructing maps
   - graphing relationships
   - deriving a formula
   - listing related ideas

6. *Problems are relevant when they are on the level of difficult examination problems.*

7. *Build new connections to previously learned material.* For example, a problem dealing with the mechanism of electrophilic aromatic substitution could be preceded by a refresher discussion on the mechanism of an electrophilic addition to a carbon-carbon double bond. Finally, these two mechanisms could be emphatically tied together by asking students to analyze the similarities and differences between the two mechanisms.

8. *Approach ideas from new perspectives.* Pairwise problem solving does that. The book gives an expert solution. How did a fellow student solve the problem?

In the section that follows, we give some selected examples of successful Workshop problems. We have made no attempt to be comprehensive; rather, we hope that each example and the accompanying text will illustrate the point that the distinctive structure of the Workshop requires distinctive problems. The problems come from the Workshop Chemistry Project books: *General Chemistry*, David K. Gosser, Victor S. Strozak, and Mark S. Cracolice; *Organic Chemistry*, J. A. Kampmeier, Pratibha Varma-Nelson, and Donald K. Wedegaertner; and *General, Organic, and Biochemistry*, Pratibha Varma-Nelson and Mark S. Cracolice.

*This Workshop problem illustrates the random round-robin method of problem solving applied to the concept of Lewis structures. An additional analysis step provides an opportunity for reflective problem solving after the Lewis structure is complete. The random round robin approach dissects a complex task into a series of discrete steps. This provides a logical sequence for problem solving; it makes the thinking process visible. Simultaneously, a formidable challenge is reduced to manageable bites. The structure helps people work together and provides experience with group problem solving. The leader plays an essential role in setting up the structure and monitoring the progress of each member of the group.*
*(D. K. Gosser)*

## Essential Concepts of Bonding

### Round-Robin Instructions

- The Workshop leader randomly assigns each group member a number between 1 and 6. If there are fewer than six students in the group, are combined steps. If there are more than six students, team steps are doubled.
- Number 1 reads the problem aloud and does the first step, explaining the answer to the group.
- When the group agrees that step one is correct, number 2 does the second step. When the group agrees that the second step is correct, number 3 does the third step, and so on. The group continues this way until everyone agrees on a final Lewis structure.
- The leader will often pose a summary question. In this case, the question is specified as the analysis step (step 7). This is a problem for the entire group. For the next structure, each person gives his or her step to the person on the right and continues the procedure outlined. Each group member should ultimately have a turn at solving all the steps.

Give Lewis structures for the molecules specified below.

**Step 1**: How many electrons are available for bonding?
**Step 2**: For every atom-to-atom connection, make a single bond (a shared electron pair).
**Step 3**: Distribute the remaining electrons available first to nonbonded pair positions (a maximum of six nonbonded pair electrons per atom, except H, which cannot have nonbonded pairs).
**Step 4**: Identify atoms that need to have an octet but do not yet have one.
**Step 5**: Rearrange electrons from nonbonded pairs of adjacent atoms to multiple bonds to satisfy the octet on atoms identified in step 4. Count the electrons to make sure that you still have the same number as you started with!
**Step 6**: Calculate the formal charge on each atom.
**Step 7**. Analysis. Are the bonds nonpolar, polar covalent, or ionic?
How many resonance structures can you draw?

**Use the Random Round-Robin to Analyze the Following Compounds:**

| | | | | |
|---|---|---|---|---|
| $C_2H_6$ | $C_2H_4$ | $C_2H_2$ | $NH_3$ | HBr |
| $H_2O_2$ | $SO_4^{2-}$ | NNO | $NO_3^-$ | $CO_3^{2-}$ |
| | $PCl_5$* | $BeCl_2$** | $C_6H_6$ | |

*Draw one structure in which P exceeds the octet rule, and one in which P does not.
**Be can have less than an octet. Can you explain why?

*Problems in stoichiometry are often reduced to substituting experimental numbers into mathematical formulas. In contrast, the structure of this Workshop requires the students to first come to agreement about the fundamental stoichiometric relationships in the combustion reaction. Once the concept is in place, it can be expressed in terms of a formula; the key point is that the formula comes from the concept and not from the book. The leader needs to manage the process to make sure that the problem evolves in this way, concept before formula. The leader can summarize the take-home points in this problem by asking the students to work together to construct a* flowchart *for the logical steps to the molecular formula.*

## Introduction to Stoichiometry

Vitamin C is an antioxidant. This class of compounds is important in biochemistry partly because antioxidants negate the effects of oxidizing substances in the body.

Combustion of a 35.5-mg sample of vitamin C produced 53.3 mg of $CO_2$ and 14.4 mg of $H_2O$. From this information we can find the empirical formula of vitamin C.

### Part I. Group Brainstorming: Setting Up the Problem

Write a chemical equation that represents the information provided. Identify the reactants and products. This equation cannot be balanced because vitamin C can be only represented as $C_xO_yH_z$, where $x$, $y$, and $z$ are the unknown integers that we are seeking.

a. Where does all the C in the $CO_2$ product come from?

b. Where does all the H in the $H_2O$ product come from?

c. Where does the oxygen in the $CO_2$ and $H_2O$ come from?

### Part II. Calculations

a. How many moles of carbon were in the vitamin C sample?

b. How many moles of hydrogen were in the vitamin C sample?

c. How many moles of oxygen were in the vitamin C sample?

d. What is the mole ratio of C:H:O in vitamin C?

e. What is the empirical formula of vitamin C?

f. The molar mass of vitamin C is 176.12 g/mol. What is the molecular formula of vitamin C?

*Model building is an essential activity for chemists. The model kit provides a way to construct a palpable representation of an abstract idea of a molecule, but the kit also structures the social interactions and provides a set of toys. Everyone likes to build. The haptic dimension of the experience immediately reveals the fundamental insight that molecules have shapes. They are not formulas; they are multidimensional objects in three-dimensional space. They have geometric properties such as bond angles, bond lengths, and even spatial arrangements of the atoms. Good models lead to the predictions of properties; that is, the properties are connected to the structure. Several examples of this structure-property relationship are explored in this problem. The leader's first role will be to guide the students to the proper use of the models. Ultimately, the leader will need to guide the students to the insight that the properties are a consequence of the structure.*

## Molecular Models

The molecular model kit consists of colored centers with connecting joints that correspond to various common geometries. For instance, carbon is black and comes with tetrahedral or trigonal planar connectors. Straws of various lengths can be used to indicate bonds. We will use the shorter lengths to indicate bonds to hydrogen, and the longer lengths to indicate bonds between any other atoms.

| | |
|---|---|
| C = black | O = red |
| H = white | N = blue |
| P = purple | S = yellow |
| Metals = silver | Halogens = green |

1. a. Working with one partner, build the following molecules: $H_2O$, $H_2S$, $CO_2$.

2. b. Based on electronegativity differences and your model, match the preceding molecules with the following dipole moments. Indicate the direction of the dipole moment. Dipole moments: 0.0 D, 0.95 D, 1.85 D.

3. Build the acetic acid ($CH_3COOH$) molecule. Draw the Lewis dot structure first. Identify the geometry around each central atom in the molecule. Does the molecule have a unique shape? What factors do you think contribute to its most stable shape?

4. Benzene is a classical example of resonance, in which all the carbons are trigonal planar, whereas each carbon in cyclohexane is tetrahedral. The molecular formula of benzene is $C_6H_6$, and that of cyclohexane is $C_6H_{12}$.

   a. Draw Lewis dot structures for each.

   b. Build models of benzene and cyclohexane.

   c. Sketch the two molecules.

   d. Using the models, build a cyclohexane with a chair shape (like a lounge chair) and another with a planar shape. Discuss the differences between the two. Which do you think is the more stable shape? Explain why you think that.

*Concept mapping is a powerful way to review and consolidate ideas that have been introduced in lecture and the text. The very nature of the mapping exercise requires students to construct the connections between concepts. This is, of course, a fundamental step in learning. The connections help us hold the ideas in place in our minds. The problem is already structured for the leader and the students. First come the concepts and then the connections. A good way to make the maps is to put concepts on "stickies." The stickies can be arranged and rearranged on the blackboard and the connections revised and restructured until a satisfying map is obtained. The kinesthetic and social dimensions of this problem are almost as good as the practical illustration of learning theory.*

## Creating a Concept Map for Chemical Bonding

A *concept map* is a diagram that shows connections between concepts. Because chemical bonding is of such overall importance to chemistry, it is useful to reflect on everything we know so far about chemical bonding. We can do this by listing all the concepts that contribute to the overall concept of chemical bonding and relating these in a concept map. A sample concept map, adapted from Herron (1996), for the ideal gas law is shown here.

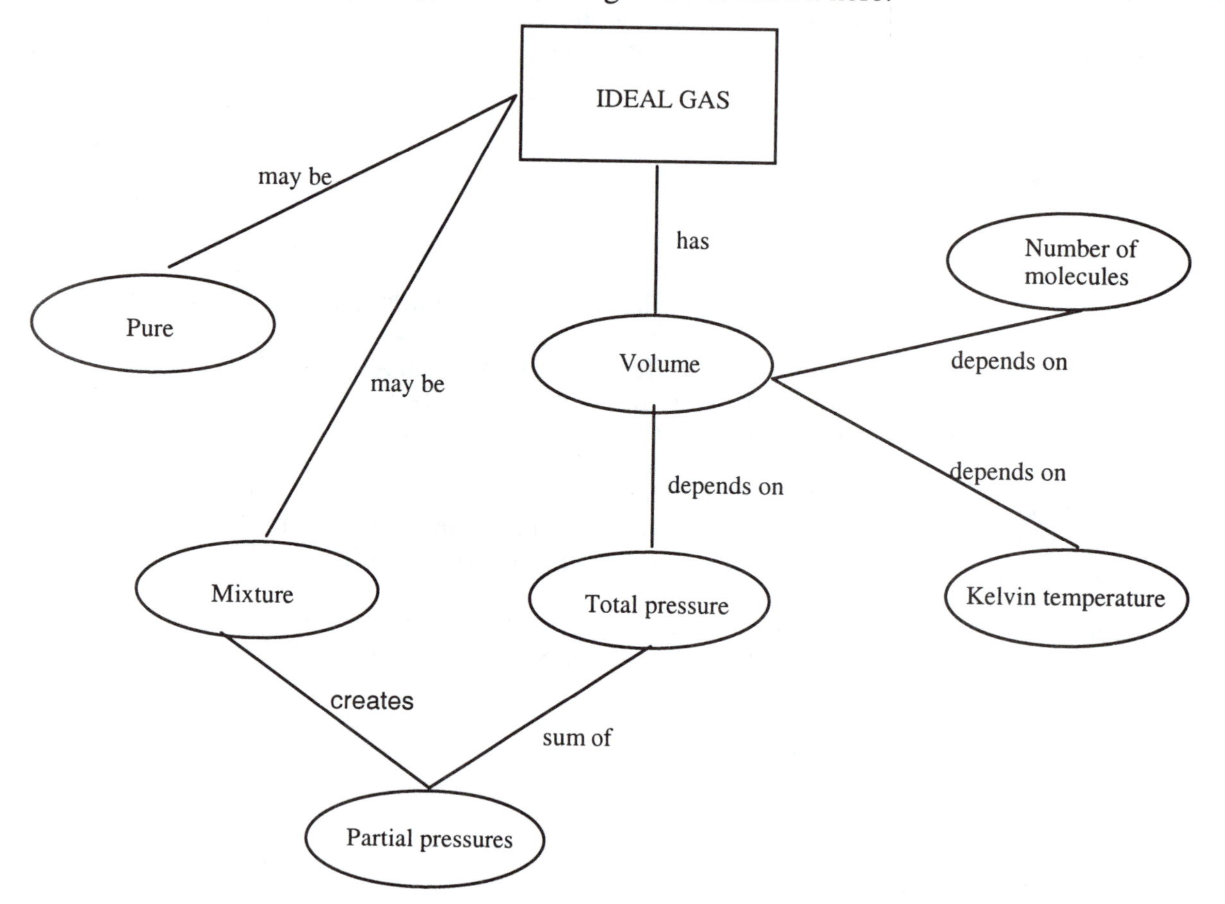


1. As a group, generate a list of eight to ten concepts that are critical to understanding the chemical bond.
2. Break up into subgroups of two to four and create concept maps using the list that the group generated.
3. Post the concept maps, and compare how the subgroups constructed their concept maps.

*A type of problem that looks like a drill problem is given next. In fact, it is an excellent Workshop problem that is built around pattern recognition skills. Fundamentally, the structure of the problem makes clear that all these seemingly diverse reactions involve the same two-step mechanism. The individual parts of the problem are easily assignable to individuals or to subgroups. The leader can use the problem as a starting point for discussions of regiochemical and stereochemical similarities and differences in these reactions. (J. A. Kampmeier)*

## Electrophilic Addition Mechanism

1. Make a table listing the electrophiles and the nucleophiles in each reaction.

2. Write a single, general mechanism that explains all these reactions

$RCH{=}CH_2 + Cl_2 \rightarrow RCHClCH_2Cl$

$RCH{=}CH_2 + Cl_2 \text{ (in } H_2O) \rightarrow RCH(OH)CH_2Cl$

$RCH{=}CH_2 + Br_2 \rightarrow RCHBrCH_2Br$

$RCH{=}CH_2 + Br_2 \text{ (in } CH_3OH) \rightarrow RCH(OCH_3)CH_2Br$

$RCH{=}CH_2 + HCl \rightarrow RCHClCH_3$

$RCH{=}CH_2 + HBr \rightarrow RCHBrCH_3$

$RCH{=}CH_2 + HI \rightarrow RCHICH_3$

$RCH{=}CH_2 + H_2SO_4 \rightarrow RCH(OSO_3H)CH_3$

$RCH{=}CH_2 + H_2SO_4 \text{ (in } H_2O) \rightarrow RCH(OH)CH_3$

$RCH{=}CH_2 + H_2SO_4 \text{ (in } CH_3OH) \rightarrow RCH(OCH_3)CH_3$

$RCH{=}CH_2 + Hg(O_2CCH_3)_2 \text{ (in } H_2O) \rightarrow RCH(OH)CH_2{-}HgO_2CCH_3$

$RCH{=}CH_2 + Hg(O_2CCF_3)_2 \text{ (in } CH_3OH) \rightarrow RCH(OCH_3)CH_2{-}HgO_2CCF_3$

*Structure proofs work very well in the Workshop setting. We have made explicit the* Observation-Deduction *approach to problem solving. The leader might ask the students to find the observations; the leader could be the scribe and record the observations in numbered sequence on the board. The leader could then solicit group discussion about the appropriate deductions from the experimental observations. There are built-in opportunities for the leader to bring out ambiguities to elicit multiple hypotheses and to distinguish between necessary and sufficient reasoning. This simple* Observation-Deduction *device provides a logical stepwise procedure that simultaneously facilitates solving the problem and engaging all members of the group. The problem ultimately requires the group to come to agreement. The procedure ultimately reveals the chain of "if-then-therefore" logic that is the source of confidence in the solution. In addition, the method gives everyone a chance to participate; even the most unprepared student can find an observation. (J. A. Kampmeier)*

## Alkenes: Reactions

1. Compound **J** has the molecular formula $C_6H_{12}$. Reaction with $H_2O/H_2SO_4$ or $BH_3{\cdot}Et_2O$, followed by reaction with $H_2O_2/OH^-$, gives the same compound **K**, $C_6H_{14}O$. Reaction of **J** with $O_3$ followed by reduction with Zn/acetic acid, or with $KMnO_4$ in acidic solution, gives only one product **L**, $C_3H_6O$. Propose structures for compounds **J**, **K**, and **L**.

**Observation** **Deduction**

**J** = **K** = **L** =

2. **M** and **N** are isomers, $C_4H_8$. **M** reacts with $H_2$ in the presence of a catalyst to give an alkane, $C_4H_{10}$. **N** also reacts with $H_2$ under the same conditions to give $C_4H_{10}$, which is different from the compound obtained from **M**. Reaction of **M** with $O_3$, followed by treatment with $Zn/H_3O^+$, gives two products, $CH_2O$ and $C_3H_6O$. **N** reacts under the same ozonolysis conditions to give one product, $C_2H_4O$. When **N** is heated with a few drops of $H_2SO_4$, it is converted to a mixture of **N**, **O**, and **P** in which **O** predominates. **N**, **O**, and **P** are all reduced with $H_2/PtO_2$ to the same alkane. Ozonolysis of **O** gives one product, $C_2H_4O$, the same product obtained from ozonolysis of **N**. Ozonolysis of **P** gives two products, $CH_2O$ and $C_3H_6O$. This $C_3H_6O$ compound from **P** is not the same as the $C_3H_6O$ compound from **M**.

**Observation** **Deduction**

**M** = **N** =

**O** = **P** =

*For reviewing a large number of synthetic reactions, the "starburst" problem works well in group settings. Every Workshop student can work on some aspect of these synthetic problems. Since multistep pathways are required and there is more than one way of doing many of the syntheses, the format leads naturally to sharing results and insights. The suggestion to work around the perimeter makes the problem open-ended. The structure of the problem provides a competitive challenge that is best met by cooperative interaction of the students. The leader can prompt, as needed; can provide a plan of attack (make C-C bonds first); can suggest new connections and functional group interchanges. The best results come when students construct their own stars. The leader can suggest starting points for new stars. (J. A. Kampmeier)*

## Alkynes: Synthesis

**See how many synthetic connections you can make on the following chart.** Starting with acetylene and compounds containing three carbons fewer, synthesize each compound in the chart. Specify reagents, conditions, and catalysts. Usually, more than one step is required. Show multiple steps by linking compounds on the chart. It will help to think "forward and backward." For more fun and challenge, work your way around the perimeter.

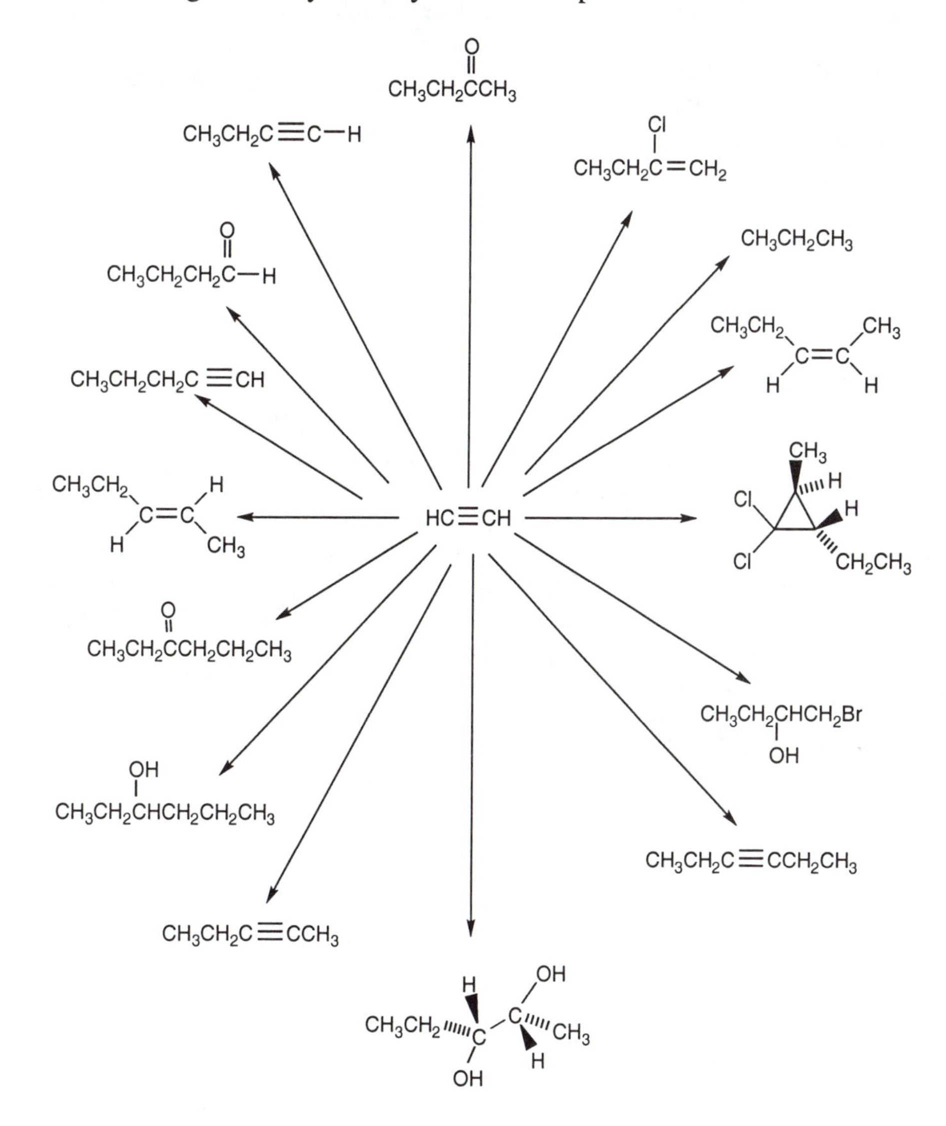

*Another type of puzzle problem is the classic "road map" illustrated here. Group participation is readily facilitated as different students contribute the products or reagents needed for different parts. The group can usually solve the entire problem in a satisfying manner, even though individual students are unsure of particular parts of the map. The structure of the problem reveals connections and relationships among a number of functional group interchanges. The road map is a different kind of concept map. (D. K. Wedegaertner)*

## Review

Give the structures of the products or reagents, as appropriate, indicated by letters **A–J**. **Clearly represent the appropriate stereochemistry.**

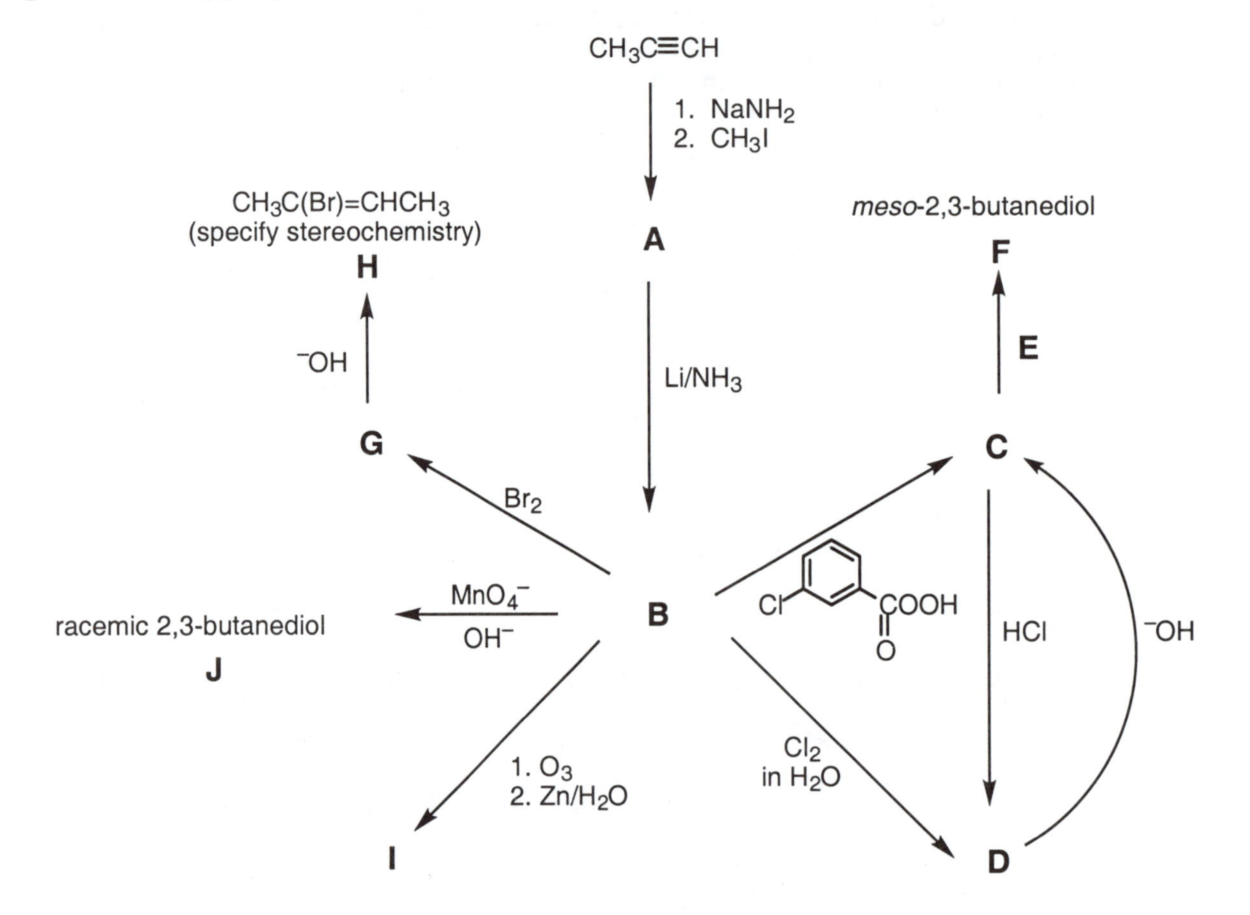

*The General-Organic-Biochemistry course provides many wonderful opportunities to explore the fundamental principle that chemical reactivity is a consequence of molecular structure. That principle is the bridge from organic chemistry to biochemistry. It is simultaneously the bridge from fundamental knowledge to practical application. We learn by constructing networks of observations and ideas; problems that require students to make connections are always productive. A group approach works well because the problems have many different parts, and students can share their knowledge of the individual parts. The leader's job is to make sure that everyone participates and that everyone sees the connections. (P. Varma-Nelson)*

## Structure: Reactivity

I. The following triglyceride can react in several different ways:

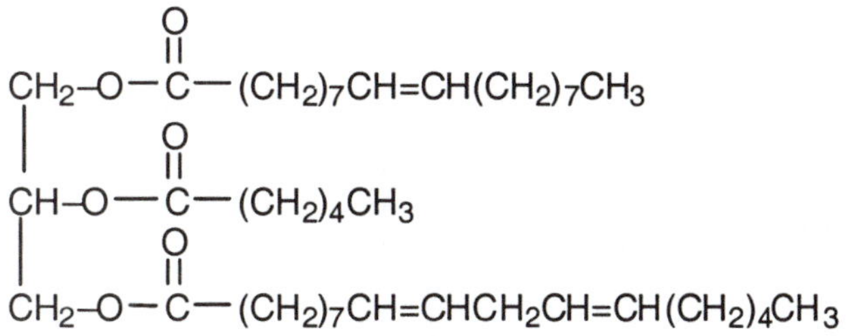

$$
\begin{array}{l}
CH_2\text{–}O\text{–}C(=O)\text{–}(CH_2)_7CH{=}CH(CH_2)_7CH_3 \\
\quad | \\
CH\text{–}O\text{–}C(=O)\text{–}(CH_2)_4CH_3 \\
\quad | \\
CH_2\text{–}O\text{–}C(=O)\text{–}(CH_2)_7CH{=}CHCH_2CH{=}CH(CH_2)_4CH_3
\end{array}
$$

A. For each of the following reactions, specify the necessary reactants, reagents and conditions for the reaction. Explain to the group which part of the triglyceride reacts and what the products will be.

1. Hydrolysis (a reaction that occurs when a fat or oil becomes rancid)
2. Saponification (used to produce soap)
3. Partial hydrogenation (used to produce margarine from vegetable oil)
4. Complete hydrogenation (to produce solid shortening from vegetable oil)
5 Oxidation (a reaction that occurs when a fat or oil becomes rancid)

B. Lipids are responsible for several features of biological membranes.

1. Explain how a high concentration of unsaturated hydrocarbon chains makes the membrane more fluid.
2. Explain why the presence of cholesterol decreases the fluidity of cell membranes
3. Explain why lipids make a cell membrane impenetrable to ionic and polar substances.
4. When lipid bilayers are disrupted they are able to reseal spontaneously. Why?

II. The structure of a protein is very complicated - so complicated that different levels of structure are specified as primary, secondary, and tertiary.

A. Work together as a group to list all the different structural elements that determine the detailed protein structure.

B. Make a diagram that shows the different levels of protein structure. Relate the various structural elements from your list to the appropriate levels of structural complexity in your diagram.

C. Consider the globular protein albumin. Discuss the effects of each of the following denaturing agents on this protein. Which covalent and noncovalent interactions will be affected? Which level of protein structure will be affected? Use drawings to explain the effect of each agent.

1. Heat

2. UV light

3. Acid

4. Base

5. Ethanol

6 Heavy-metal cations such as $Pb^{2+}$ and $Hg^{2+}$

7. $H_2N-C(=O)-NH_2$ (urea)

*Most presentations of chemical kinetics start out with the definition that the speed of a chemical reaction corresponds to a change in concentration per unit time. This idea is easily digested because of the analogy to familiar ideas about the speed of moving objects such as trains, cars, and bullets. The next step in the discussion usually builds on qualitative observations about the rates of chemical reactions: "heat it up and it goes faster" or "increase the concentration and it goes faster." Then, suddenly, an equation appears: rate =* k*[reactant]*$^n$. *The following Workshop problem uses a simulation to bring an intuitive feel to the study of chemical kinetics. Like many good Workshop problems, this one includes something physical and concrete to do; there is a kinesthetic dimension to the problem. There is also something fun to do: a game is played with other people. The game is not just a game, of course. It is an experiment in which the students measure (count) concentrations as a function of time. Ultimately, the observable experiences are transformed into analytical descriptions (graphs) of the relationships of concentration to time in the different games. (D. K. Gosser)*

## Chemical Kinetics

***Work in groups of threes, with two players and one scorekeeper.***

1. Consider a simple chemical reaction, $A \rightarrow B$, that follows a first-order rate law, rate = $k$[A]. You will model this reaction with pennies. Start with 100 pennies, which will represent the initial concentration of A, 100 mM. Each penny will therefore represent 1 mM. It may be useful to mix in higher denomination coins such as quarters, nickels, and dimes to make the counting easier.

   Student A (SA) will represent the concentration of A, and student B (SB) will represent the concentration of B. We will represent the reaction of A to form B by passing pennies from SA to SB. Each exchange of pennies will represent one second of time. Student C will record the observed results.

   We will model a reaction in which 10% of the concentration of A reacts per second. Thus for each second of time (exchange step), SA should transfer 10% of his/her pennies to SB. Round fractions to the nearest penny. Continue this exchange for 15 seconds.

   a. Record concentrations of A and B (number of pennies) each second (after each exchange step) in the following table.

| Time (s) | [A] (mM) | [B] (mM) |
|---|---|---|
| 0 | 100 | 0 |
| 1 | 90 | 10 |
| 3 | 81 | 19 |
| 4 | | |
| 5 | | |
| 6 | | |
| 7 | | |
| 8 | | |
| 9 | | |
| 10 | | |
| 11 | | |
| 12 | | |
| 13 | | |
| 14 | | |
| 15 | | |

   b. Plot the concentration of A versus time on a graph. Use a different color to plot the concentration of B versus time on the same graph.

2. Let's apply the modeling technique developed in question 1 to a reversible reaction,

$$A \rightleftharpoons B$$

In each second of time (exchange step), allow 10% of A to react to form B, and allow 10% of B to react to form A.

a. Change roles.

b. Record the results in a table like the one in question 1.

c. Plot the concentration of A versus time for the 10%/10% reaction on a graph. Use a different color to plot the concentration of B versus time on the same graph. Compare this graph with the one of the irreversible reaction A → B in question 1.

3. Try the reversible reaction in which 10% of A reacts to form B, but only 5% of B reacts to form A. Record the results in a table and plot the concentration of A versus time for the 10%/5% reaction on a graph. Use a different color to plot the concentration of B versus time on the same graph. Compare this reversible reaction with the 10%/10% reversible case in question 2.

# *Chapter Four*
# *Workshop Leader Training*

**Vicki Roth, University of Rochester**
**Mark S. Cracolice, The University of Montana**
**Ellen Goldstein, City College of New York**
**Vivian Snyder, University of the Pacific**

> *There was a point at which I just really had to smile. Student P was up on the board trying to work out a problem, and he was stuck, so some of the kids started shouting out suggestions. Student P was still stumped, so he turned to Student O and said, "Man, I still don't get it, show me," and he tossed him the chalk. It was incredible because I wasn't even part of the equation. They were looking to themselves for the answers and explanations, not me. It was exactly what I had pictured the Workshop as being like.* *Bob Tubbs, University of Rochester leader*

To help you find your way through this chapter, here is a directory for the following pages:

## Introduction

Our day-to-day experiences in the Workshop Project have taught us that the strength of this model lies in the character and caliber of the interactions among Workshop participants and with their group leaders. We have learned that good peer-led education does not always spontaneously emerge when people are simply grouped in a room, assigned an agenda, and given a set of problems to solve. As you know from your work with students in the classroom and with your colleagues on committees, nudging people toward the formation of strong, consistent work teams is a skill and an art form; it takes practice to learn how to do this well.

New Workshop leaders face real challenges: the content we ask them to deal with is complex, the students who attend their Workshops will not all see the advantages of working as a team at the outset, and the leaders' group leadership experience is often the first substantial professional responsibility they have assumed.

Despite these obvious challenges, it is tempting to downplay the need for training. We could presume that new leaders, by virtue of their own prior academic accomplishments, will enter our Workshop programs with sturdy leadership skills in place. But this is not a dependable assumption. We all know instructors whose personal command of the discipline does not translate into adept guidance of others through the material; the same can be true for student leaders. Second, this model asks new leaders to make a basic shift in the way they have commonly experienced the academic world. They may have had some prior involvement with study groups and Workshops, but it is likely that much of their school time has been in passive, not in participatory, modes of instruction. Essentially, we are asking our new leaders to do as we say rather than what they have been shown throughout most of their school years. It is vital to the success of the program that leaders make a shift in orientation by learning and practicing new skills.

If you are thinking about establishing a leader training program, you and your team will encounter a series of questions, investigations, and choices to make along the way. The rest of this section is intended to walk you through the most crucial of these choices in the order you will most likely face them, starting with decisions about the basic structure of the program and ending with budget matters.

## Getting Started: Setting Up the Program

***Decision One: Structure of the training program***. The first round of choices relates to the level of training most appropriate for a specific institution. The Workshop training model has proved to be a flexible one, with arrangements that can be grouped into three types or options for programming, from more to less intensive. Before you make decisions about how comprehensive your training program should be, we encourage you to survey the resources and climate on your own campus. The decision should be based on a real assessment, not just general impressions, since there may be more support available locally than is apparent at first.

*Option One*, which can be a good fit for smaller Workshop projects, involves a series of meetings between the Workshop leaders and the instructor. These sessions are like the informal staff meetings instructors typically have with teaching assistants. This approach can include opportunities for the leaders and the instructor to debrief about the past week's Workshop sessions and to conduct practice runs of the next modules together.

At this level, some programs have included written job descriptions for the leaders; if so, participation in these meetings may be stipulated in leaders' contracts. Option One may include the expectation that leaders read about pedagogical issues and keep a journal about their experiences in the Workshop sessions. Formal academic responsibilities, like the completion of a paper or project, however, are usually not part of the agreement. Much of the discussion during these sessions tends to be about the content of the modules themselves, with some opportunity to discuss learning theory and group dynamics.

*Option Two* is made up of series of training meetings held throughout the term, combining the resources of the science faculty and campus learning specialists. Again, like Option One, the training program may be included in leaders' contracts. The knowledge base of learning specialists ensures more attention to cognitive theory, pedagogical issues, and group dynamics and offers the possibility of directly applying learning theory to teaching specific principles and ideas in the science course itself.

On your campus, a viable connection may currently exist between science faculty and learning specialists; if so, Options Two and Three serve as natural extensions of this working relationship - and perhaps the shared sense of mission - that already is in place. Other programs may need some lead time to cultivate a connection between the two programs. In fact, science faculty initially may find that the varied administrative placement of learning specialists across different types of colleges and universities makes it difficult to establish a contact. Sometimes a little investigative work is required.

Many learning specialists, as you might expect, are employed in *learning centers*. These programs may be located in the academic side of the house, or, alternatively, in student affairs divisions. Some centers have traditionally functioned as "fix-it" shops, whose prominent role on campus is the rescue of students in trouble via tutoring and counseling. Others are also directly involved in faculty development and curricular reform and may employ staff members who specialize in science education. The latter sort, naturally, tends to be the most resourceful in terms of supporting Workshops, but even the former may be able to provide help for a leader training program, given that many centers are involved in ongoing tutor training.

The personnel employed in learning centers vary considerably. Some centers may be overseen by faculty members or professional staff who have considerable long-term investment in their programs; others may be staffed by adjunct faculty or graduate students who may appear less experienced in the sort of curricular development work under discussion here. We have learned to avoid drawing hasty conclusions about who can offer the best help, however, since valuable expertise for Workshop programs is often provided by part-time faculty, staff, and graduate student assistants.

Other obvious sources of learning specialists are *departments of education*. Faculty and graduate students in math and science education might be the first people to consult. Even though their emphasis tends to be on K-12 issues, much of what they have to say about learning theory and pedagogical strategies can be extended meaningfully to the college level. Other specialists, perhaps in education departments that focus on faculty or curriculum development, may also be of real help. Sometimes these departments are eager to find placements for graduate or undergraduate student interns who are willing to work for academic credit. These arrangements can be of mutual benefit: the interns gain first-hand, real-world experience, and your program can profit from their free expertise.

Your campus may have a *faculty development center*. If so, this is yet another promising source of support for a leader training program. The formats and missions of teaching centers vary widely across the country; a phone call or two should clarify if your center can assist your Workshop program.

In a slightly different vein, you may also want to investigate the potential contributions of your *library staff*. Not only can these staff members help you locate information about training and pedagogy but they also may be able to provide specialized help for your leaders in finding additional information about group facilitation and science education.

*Option Three* is a training program based on a credit-bearing training course, often taught collaboratively by science faculty and learning specialists. Positioning the leader support within a "real course" provides a structure and a set of built-in expectations that encourage more intense engagement with the training process. The course allows for regular, frequent contact between the leaders and the faculty, and it also offers the opportunity for more deliberate investigation of all sorts of training issues - those that relate to learning theory as well as those that are science-content specific. Our goal has been to develop interactive seminars within this training option, rather than lecture courses we know that if we are to reinforce the idea of student-centered learning, it is important that we do more than talk at our students about this model.

Team teaching this course sends a powerful message to the leaders. Meeting with both the science instructor and the learning specialist during the class sessions endorses the value of both content and theory and sets up a dynamic round of discussions about both that would be hard to duplicate in separate conversations. As an example, the science instructor may introduce the concept of three-dimensional molecular geometry, with the goal of explaining how the distribution of electrons within the structure leads to predictable molecular shapes. The learning specialist can use this opportunity to discuss the differences between concrete and formal learners and how these differences apply to students' mental models of three-dimensional molecules. A discussion can follow in which the science instructor and the learning specialist work together, with input from the leaders, on how to guide students through molecular geometry problems.

Because this is an academic course, we can ask students to complete papers or projects that reinforce the objectives of the Workshop model. These reports have in practice turned out to contribute real assistance to the project. This aspect of the training program is spelled out in some detail in the section "The Training Itself." More about the content and direction of these options is outlined below, but before we move on to the next choices, it is worth noting that decisions about the intensity of training program are not "one time only" choices. In fact, most good programs tend to evolve across time to suit the needs of individual sites.

An additional note about Option Three: the infrastructure established by a strong team-taught leadership course can help with the "lateralization" of a Workshop program. Other faculty in your institution may be more enthusiastic about trying Workshops in their courses if they see that an effective support system is in place.

Regardless of the structure of the leader training program, the connections among faculty and leaders are central. We know that the incentive for attending and participating in training may appear to be contractual, but leaders show up for these meetings when their relationship with the faculty is meaningful to them. Crucial to these connections are the ways in which faculty members serve as models during the reviews of upcoming Workshops with the leaders. When faculty serve as good Workshop leaders during training sessions, student leaders are able to integrate both the material and the processes involved in guiding others.

***Decision Two: Finding your leaders.*** A common belief among those starting a Workshop program is that good leaders are hard to come by, and this certainly seems true during the first year of a program, when no natural pool of candidates has had time to form. It is hard to believe at first that potential leaders will be knocking at your door in a semester or two.

While you are waiting for a leader pool to emerge, direct recruiting is probably the best strategy. Look for those talented students in your courses who demonstrate a reliable, committed approach to their own work. It is highly flattering to most students to receive an invitation from a professor to be part of a new initiative. Being asked to become a Workshop leader is, in fact, a mark of accomplishment and merit. Marketing it as such is an honest, effective recruiting tactic. A position as a Workshop leader can truthfully be "sold" as a way to develop leadership skills, to gain knowledge of group and collaborative learning, to better understand the course content, to improve one's own learning, and to enhance future career and graduate school opportunities. When the Workshop program has been in progress for a period of time, a set of successful Workshop students starts to accumulate. These capable participants then become an obvious source of potential leaders for subsequent semesters.

What follows next are factors that Workshop Project partners have used to develop a pool of strong applicants. After your own Workshop program has been in place for a term or two, the most salient selection factors for your institution, tend to become more apparent. This discussion about *what* to look for is followed by a description of *how* to evaluate these qualities in your candidates (**Decision Four: The selection process**).

*Recent completion of the course.* Given that there are no fixed boundaries governing the content of most of our science courses, it helps to hire new leaders who have taken the course in question during the previous academic term or two. This flattens the learning curve, as these leaders then are familiar with the course format, subject matter, and performance expectations.

*Successful completion of the course.* In general, a Workshop project is best served by hiring very academically talented students. Our "A" students form a obvious pool, since they tend to be good performers in all academic areas and adept at negotiating the social and political terrain on our campuses. A caution should be noted, though, about those highly gifted students who find the study of science, at least at this level, to be nearly effortless; they may not always be the most empathetic coaches for students who struggle with the material. On the other hand, their considerable insight may help them become skilled guides for their somewhat less talented peers.

A tentatively enthusiastic note should be sounded for including slightly less successful applicants in the applicant pool; "B" students may be able to bring to their groups an understanding about the learning process that "first tier" students may not have. A particularly promising group of leader candidates are those students who did not succeed in their first attempt at the course but triumphed during a subsequent enrollment. Not only do they recognize the minefields in the course but they can also transmit a convincing message about perseverance to their own group members.

*Responsibility.* Another strong indicator of future success is the commitment candidates demonstrate toward their work overall. Brilliant but somewhat flighty students who sail into class late, turn in reports a week past their due date, and forget to show up for the final exam are likely to bring this same confusion to their role as leaders as well, forcing you, as the program director, into a babysitting role. We give the nod to those who can be counted on.

*Of course, a peer leader should have a good working knowledge of chemistry, but leadership and interpersonal skills are just as important.* *Chris Boeschel, St. Xavier University leader*

*Leadership potential.* A talent for guiding others is clearly an important selection criterion for a Workshop program, but it is also one of the hardest factors to assess in incoming applicants. On campuses with typically small classes, the course instructor and/or colleagues in the department may have been able to observe the group skills of some of the applicants over the course of a previous semester, but in many introductory science courses, the large number of students reduces the opportunity to judge these skills. It is hard to spot leadership potential in a lecture hall filled with rows of students.

After a Workshop program has been established, however, a natural mechanism develops that promotes this sort of long-term observation. The Workshop sessions provide many hours of real-world discussions. Thus, group leaders who have attended well to the dynamics of their students' interactions can be excellent sources of information about the best recruits for subsequent academic terms.

There is no one type or personality that makes for a good leader; in fact, program directors should be cautioned against creating an army of clones. Not only does a sameness reduce the aggregate creative energies of the leaders, but it also diminishes the likelihood that the needs of different types of learners will be reflected in the instructional approaches developed in the program. That said, there are certain leadership qualities to look for, however diversely they may be expressed by different candidates. Paramount among these is evidence of genuine interest in others. It is easy to dismiss those candidates who are openly condescending toward their peers; it is harder to watch out for those "lone wolf" students who have a more subtly paternalistic attitude toward others. These folks will have a harder time seeing the need for real group interaction during their Workshop sessions. Their tendency will be to emulate a lecturer's approach to the group, subverting the goals of the model. In contrast are those applicants who, by virtue of personality and experience, enjoy tuning in to those around them. It is a bonus if their attentiveness to others is focused not only on general social exchange but also on the academic issues at hand.

A necessary condition for developing this talent is the ability to listen. Most of the features marking true attentiveness to others are fairly subtle but still observable. Frequency of eye contact, for example, is a good marker of mindfulness, as are those small confirming remarks and gestures that people interject in order to encourage individual contributions and the overall progress of the group as a whole. An applicant who is able to pay attention while someone else is speaking, and nod encouragement, is likely to be a better leader than those candidates who can scarcely wait for their turn to hold forth in front of the group.

A more indirect quality to look for is the ability to understand how someone else is positioned in relation to new information. The leader who knows what a student is ready to learn next is a keeper. Evidence of this quality, or the lack of it, often shows up in phrases or comments leaders make in response to students' attempts at Workshop problems. However well intentioned, some "motivating strategies" are founded on an underground form of intimidation. The directions from less skilled leaders may coerce students into pursuing a train of thought because they feel they are expected to do so or because they feel threatened, however inadvertently, by the sophistication of the leader's grasp of the material. The best leaders know how to offer guidance that encourages others to own the problem, and ultimately, the solution. It should be said, though, that it is a rare leader who has this ability fully developed at the outset; most of us take a professional lifetime to hone the talent of skilled guidance.

***Decision Three: The selection process.*** In this section, we switch gears, moving from a discussion of abstract selection criteria to the more concrete logistics of recruiting and interviewing candidates for Workshop leader positions. An effective start is with an interest meeting. We recommend holding this session midway through the academic term before the start of the new group of Workshops, for example, scheduling a meeting in March to recruit new leaders for the following September. After generating a list of our most plausible candidates (based on grades and input from the Workshop leader and the course instructor), we send out invitations to the interest meeting, congratulating the candidates on their previous success in the course and emphasizing the honor inherent in the invitation. A sample invitation and other forms related to the hiring process are included in Appendix II.

The session itself can be very informal; this is a time for people to get together and talk about the program and perhaps to distribute printed material about the Workshop model. It is important for several reasons to involve previous Workshop leaders at this point: first, because they can transmit information about being a leader that we, as directors, can know only indirectly; secondly, it also is reassuring to potential newcomers to hear from experienced leaders that this work is do-able and personally worthwhile. If we are to make our endorsement of student-led learning stick, we need to integrate student input from the very beginning of our leaders' contact with us. By doing so, we begin the training of potential leaders in the model even before the application process has been completed.

At this meeting another important endorsement is the connection of the science faculty and the learning specialists, if your training program involves this kind of collaboration. Student leaders naturally tend to put the most stock in what their science instructors have to say because they are likely to be majors in these departments, and they often have established some kind of prior working relationship with their science instructors. The content and direction of cognitive theory, however, is often fairly unfamiliar terrain for most potential leaders. Even worse, some new recruits may have fallen into the "no training needed" assumption mentioned earlier; they know they did well in the course, and probably in other science courses as well, so it is easy for them to believe that installing this science proficiency in others requires only a series of "information transfers," or, in Freire's (1971) terms, "banking" operations. Tip-offs that this assumption has been swallowed whole are comments from new recruits that describe the upcoming training topics as "just common sense" and "obvious." A joint description of the program presented by the science instructor and the learning specialist can help them see that participating in leader training is a valid way to further their own professional development.

The interest meeting is also the place for nuts-and-bolts information about the program. A Workshop leader job description can be distributed and discussed, and the interview process can be outlined. We like to distribute job applications during this meeting so that the newcomers walk out with something tangible to do as a result of this session. Appendix II includes a sample job description and application.

Although the interest meeting is an ideal moment in which to distribute these applications, we recommend that students fill them out and return this paperwork at a later specified date. They need a cooling-off period to gauge their own level of commitment, and you will want to ascertain that they are the sort of potential employees who keep their promises and take care of business. Also rapid-fire responses to questions like Why do you want to be a Workshop leader for this course? and What qualifies you to be a Workshop leader? are likely to be cliched and meaningless. Applicants need time to prepare a thoughtful response.

When students turn in their completed applications, they can sign up for their interviews. We have found group interviews to be an ideal format, since they offer the opportunity to observe interplay among prospective leaders. We like to provide scenarios of typical Workshop situations as jumping-off points for discussion. Here are a couple of our favorites:

> This is your first meeting as a study group leader, so you are still a little nervous. Wouldn't you know it, one of the first things that happens in your session is that a student asks you a question you can't answer. All eyes turn to you. How will you handle this?
>
> You have seven students in your Workshop, one of whom monopolizes the group. The other students are getting annoyed. What will you do about this?

As the group works its way through a series of these not-so-pretend situations while we observe from the periphery of the room, differences among personality traits and levels of social dexterity become evident. Some students plunge right into the discussion; others hold back until they feel more self-assured. You will see how some students want to take center stage as much as possible, whereas others are self-effacing to the point of invisibility. You are looking, of course, for the ideal balance, the student who can take charge of a line of discussion by drawing out the opinions of the other participants. An amiable sense of humor and an ability to maintain eye contact with other members of the group are also marks of good leadership potential. To find and track these characteristics, it helps to have more than one observer in the room during these interviews. Again, this is a time for the science faculty and the learning center staff to work together, especially if both are to be part of the training process. After the interviews are over and you have begun your deliberations, it helps to review your previously agreed-upon selection criteria and job description. As you make your hiring decisions it may be important for you to consider the balance of gender and ethnicity within your final slate of candidates. A group of leaders that is noticeably top heavy in one demographic category or another is likely to send an inadvertent message to Workshop participants.

Once decisions have been made and offers have been tendered, one more meeting, individually or in groups, can help. Officially, this is the time to distribute letters of agreement and/or sign contracts (see Appendix II for a sample contract). Unofficially, this meeting offers us the opportunity to check one last time the sincerity of each of the new leaders' interest in the program, and it gives them a chance to ask the sort of questions that many applicants may be unwilling to bring up during the interview process. This "sealing of the deal" is especially important when leaders are being hired at the end of a spring term for the following fall. A summer can bring many changes in students' lives; we want to be sure that the Workshop program has evolved as a priority for them before they leave campus.

*I think you will find the Workshop full of excitement.*

*Ren-yu Zhang, University of Rochester super leader*

***Decision Four: Finding your program assistants.*** When we began our Workshops, few of us envisioned the need for student assistants to our programs. Luckily for us, a few leaders recognized our need for a hand with the daily operations of our programs and stepped forward early. Over time, the role of these program assistants (or super leaders) has become more formalized. They often help out with finding rooms, keeping track of attendance, getting materials together, observing Workshop sessions, reading the journals, and, most importantly, providing the voice of experience for newcomers to the program.

Perhaps this section is mistitled, for we have not "found" these leaders. This is a self-selecting group, so they tend to find us. So watch for the leaders who spend time hanging around your office, asking questions, and offering to help; they may waiting for an invitation to take on more responsibility in your program. Their assistance can be invaluable to you, and this experience often proves to be one of the defining experiences of their college careers. Their participation in this role then adds an entire tier to the Workshop model, adding greater emphasis to the peer-led component of the program.

***Decision Five: Timing your training program.*** As with many decisions about the program, the scheduling of leader training needs to be a good fit for specific institutions. Regardless of the specific format of the subsequent leader training program, many have found that a retreat before the beginning of the term is an excellent way to begin. We are frequently asked if training should be scheduled for the term prior to the one in which they begin their duties. In general, we have found this not to be an ideal choice. The loss of immediacy makes most of the discussion about Workshops dry and abstract. Given the fluidity of our student populations, investing a great deal of effort in training people who do not turn out to be available during the subsequent term is wastes resources and energy. Starting the training very close to the first day of Workshops is a little scary for all concerned, but we have discovered that the high motivation of the leaders at that point creates for many "teachable moments" during the training sessions.

***Decision Six: The training program budget.*** The pedagogical issues related to the Workshop model, as intriguing as they are, cannot be attended to if the financial infrastructure does not hold together. Leader training does not have to be an expensive component of the Workshop program, but there are some expenditures to be accounted for in all the options listed. A detailed discussion about funding Workshops is outlined in Chapter 5, "Institutionalizing the Workshops"; a brief summary of the typical costs of a leader training program follows.

*The staff and faculty time devoted to conducting training.* These costs vary widely, depending on the choices that were made about the type of training program to be established. Staff and faculty costs may include release time, stipends for student assistants, and salary reimbursements for learning specialists.

*Clerical support and materials.* These costs are modest, since few if any specialized materials are required for training. Some secretarial support and the cost of copying, printing, and books needs to be included in the budget.

*Computer support.* Most programs, especially larger ones, need access to computer facilities that allow for easy compilation of attendance and student performance data, and as our campuses have become more "wired" we have found web-based communication with our leaders to be an effective and efficient way of handling all the little queries and glitches that show up in the daily operation of a program. Often it is possible to piggyback on existing computer resources on campus. If this is not an option, then these costs should be incorporated into your budget planning.

*Through taking the Workshop leader class, I have learned more about myself as a person, student, and a Workshop leader. In short, I really believe that I have grown and have benefited through the Workshop experience.* *Danny Mui, CCNY leader*

## The Training Itself

***First things first: the first few days of training.*** The end result of a successful recruiting and selection process is a roomful of enthusiastic new leaders waiting for you on the initial day of training. The time to begin this training needs to be tailored to individual programs. Each campus has its own schedule and rhythm. The goal is to make the training program mesh well with the way things typically work at your institution. What follows here is a menu of what could be included in leader training. This is a flexible guide; as no training program, whether it be modeled after options one, two, or three outlined in the preceding section, can incorporate everything described here. The goal is to offer suggestions and encourage you to choose the elements most pertinent to your own program.

*A retreat.* This is an effective way to set up a preterm orientation. Having a whole day or two with a new group gives you enough time to take care of all the inevitable fine-tuning of logistics that is required at the start of any new academic term. More importantly, a retreat provides a substantial opportunity to develop cohesion among the leaders, always one of the most essential elements of our Workshop programs.

[One caution: Many TA training programs around the country are one-shot deals; that is, pedagogical directions and advice are bundled completely into a one- or two-day session with little or no continuing follow-up. Becoming a good leader is a developmental process, so we would encourage you to avoid a one-time-only training format.]

*Breaking the ice.* If you are conducting a retreat, it is a useful investment of time to begin the session with refreshments and some get-acquainted activities, despite how eager you may be to get down to the business of the Workshops themselves. Most ice-breaking strategies such as asking students to call out names as they toss balls to one another, are fairly silly; that is why they work. The goal is to break down our usual script in the classroom; that is, have students walk into the room, take a seat, and wait politely for the authority figure to tell them what to do. We want our leaders to feel comfortable with new approaches to the classroom so it is best to start off our training with something other than having them sit in rows. To help you get started with this, we have included few sure-fire icebreakers in Appendix II.

*Getting down to work.* When it is time to turn the group's attention to the Workshops, invite your students to tell you and one another about their intentions for their groups this term. This discussion helps them focus on the work before them, and it allows the trainers to spot inappropriate or impossible goals, note these as topics of future discussion, and plan for future redirection. For example, some enthusiastic new leaders are likely to say that they want their entire group to get A's in the course, or that they want to be available as much as possible for their students. Comments like these are cues for the program directors. They know that two of the goals for the retreat will be to help these leaders set boundaries for themselves and develop strategies for dealing with the non-A students they are sure to encounter.

> *I was very nervous before the class. Last night I woke up in the middle of the night due to a bad dream about orbitals. I had to check my book right away.*
>
> *Heidi Phillips, University of Rochester leader*

In addition to talking through their goals, this is also the time for the leaders to hash through their worries about their new responsibilities. Now that the slightly heady rush of being recruited and selected for this position is over, they begin to recognize the challenge of getting their own groups off to a good start. Most of them have a few qualms about this, and the emotionally honest ones among the group will be willing to describe these concerns in public. With a little strategizing, this discussion can be converted into an effective modeling of the Workshop approach itself. Teams of new leaders can be asked to turn their list of concerns into a chart (have sketch paper and markers ready). A chart produced by novice leaders at this point in the training process might look something like this:

**Things That Worry Me**

- Being asked something I can't answer
- Forgetting how to solve some of the Workshop problems
- Not being able to explain things very well
- Dealing with really shy students
- Putting up with rude students
- Filling up the time
- Running out of time
- Being too busy to do a good job
- Working with students who aren't all at the same level
- Not being liked by my students
- Not knowing where my Workshop room is
- Not knowing if the room will be equipped well (blackboards, chalk, good desks)
- Not knowing how to keep track of attendance
- Not knowing how to get the group started on the first day

Once the charts have been completed, ask a spokesperson from each team to post their pages on the wall and briefly describe them to the larger group. These admissions can be of real value to the group as a whole. For one, a public discussion tends to defuse much of the tension around tough issues; new leaders quickly recognize that their concerns are shared by others. Secondly, this sort of activity sets up the expectation that the training sessions are a safe and welcoming place for the airing of troubles. But most importantly, as you will witness, the group tries to help itself with these potential problems, thus asserting authority and taking creative control in their own hands.

The trainers - or, better yet, some of the experienced leaders - can describe how to maintain attendance records, how to handle any grading that might be part of their duties, how to locate their Workshop rooms, and the like. Leaders can visualize their upcoming responsibilities better when they have been provided with this nuts-and-bolts information.

Leaders also need to know what you expect of them in the training program. A credit-bearing training course is likely to generate a syllabus and a course schedule (see samples in Appendix II). But even if your training program does not carry course credit, tell your leaders about your expectations regarding attendance and participation in the training meetings, completion of reading assignments, and the keeping of a response journal.

*The reflective journal.* The idea of a response log merits thoughtful consideration. We have found these journals to be our single best assessment tool for the program, for they generate immediate and detailed records about each Workshop session held during the semester. The description of the journal assignment might sound something like this:

> We'd like you to keep a journal of your Workshop experiences this semester because we need your feedback. So, before your first session, buy a composition notebook and bring it with you on your Workshop day. As soon as you can catch a moment after your Workshop is over, sit down and write us a page or so describing what happened during that session and how you felt about it. Then drop it off - within 24 hours of your Workshop, please - and we'll take a look at what you had to say and write you a note in return. We'll be sure to get your journals back to you before your next session.

These journals turn out to be a gold mine. Although "debriefing time" belongs in the training meetings themselves, an in-class discussion cannot cover everyone's detailed reports, especially in our larger programs, where there is simply not enough time for everyone to speak in detail. Also, the training meeting often is not the right format for downloading more specific concerns about individual students and the leaders' worries about their leadership competence. The journals provide a venue for more private discussions. The leaders may not recognize at the start the need for a log - as science students, they may not be used to writing much yet - but by the end of the first session, they usually have something they want to tell you. As noted above, our length requirement has been modest, but most leaders write more pages per week by choice. So you will read about the Workshop problems that seemed especially effective and the ones that did not. You will find out about the Workshop participants who seem to be keeping up in the course and the ones who are faltering, and you will learn about the confidence level of each of your leaders. Our written responses to their comments probably help - or so they tell us - but, in truth, the real power of the journals is the articulation of the leaders' own thoughts. It is common to watch them work out problems for themselves within these journal entries.

*As for the Workshop itself, I think things ran generally well today, although I'm concerned that I'm "teaching" too much and not letting the students figure everything out. A few things I realized after the first sessions are that ... I need to hold back a little with my interference in the Workshop; I'm not expected to be the world's greatest Workshop leader on my very first try. I'm sure I will improve as time goes by.* *Melissa Langdon, University of Rochester leader*

*The first minutes of the first Workshop.* Group leaders are also eager to learn some *opening-day strategies*. You will want to help them feel equipped to face the group during those first few minutes. The Workshop problems themselves provide a structure for discussion, but they need a few warm-up tactics or two to help them get started. Be sure they intend to introduce all the Workshop members to each other and have a few ice-breaking exercises in mind. Then, when it is time for them to focus the group's attention on the problems of the day, they may want to use a specific mechanism to structure the interaction among the students. For example, the leader can set up a round robin technique for problem solving (see Chapter 3, "Writing Workshop Materials," for more on this) or break the group into two or three smaller teams, each of which can develop a group answer for a portion of the question. The teams' answers are then shared with the whole group for discussion.

It is important for the leaders to feel comfortable with the content of the modules. Good confidence builders are *practice Workshops*, held in the training class itself. Leaders can be guided through the Workshop problems by the instructor. In this way, the faculty member can help the students review the material and model the techniques and strategies of good group leadership.

More direct practice in leadership can be provided by asking the student leaders to form small groups, with one member of each group serving as the leader for that day and, importantly, another member acting as an observer. After the group works its way through a couple of problems, stop the teams and allow the observers to talk to their colleagues about what they saw during the session. These observations can focus on the performance of the leader and also on the functioning of the group itself. Here is a short list of questions the observers may wish to address.

- How easy was it for the group to get going? How did the leader help start the discussion?
- How well did the group dissect the question before plunging in?
- Did everyone contribute? If so, what helped make this happen? If not, what prevented full participation?
- Did the group focus exclusively on getting the answer to these problems? Or did they pay attention at any point to the important concepts within them?

As the teams of group leaders think about this feedback they are sure to be doing some mental rehearsing for their future sessions with their own students. The ultimate goal is for the leaders to see this exercise as so worthwhile that they form leader teams of their own to practice for the upcoming Workshop units. Some programs have been able to find a spare room in the department that can be designated for this purpose. A room like this is a great place for the leaders' mailboxes, a table and chairs, and a blackboard. Establishing a location for the leaders to work together helps them form a cohort.

***Once the training is underway.*** After the leaders have held their first Workshop session, issues that may previously have seemed somewhat abstract now take on a more immediate character, so the second training session needs to focus largely on debriefing. A general question like, Well, what happened during your Workshops? is likely to elicit considerable discussion. As experienced faculty and staff, we know that the best-laid plans of instruction seldom follow a straight line and that good leadership often means responding to the teachable moments that present themselves, whether or not they were in our lesson plans.

Because our group leaders do not yet have years of experience and the perspective that this bestows so they may feel disheartened to find that their students fail to coalesce immediately into dynamic work teams and, in fact, may not seem initially enthusiastic about tackling these Workshop problems at all. As these worries surface in the training class another solid opportunity arises to demonstrate how students can solve their own problems. With only the most gentle nudging (Did anyone else run into this same problem? How did you handle it? Any other ideas come to mind?), other leaders begin to offer their perspectives and possible strategies for the next Workshop session.

*I encountered obstacles while trying to enhance group discussion and problem solving, e.g., withdrawal or separation of an individual. The leader training course seemed like a manager's conference. As we met we were able to discuss techniques we used, thus sharing new ideas. Group interactions are smoothly improving as members are getting accustomed to their peers' helping to solve their problems instead of their Workshop leader.*

*Makris Anastasia, CCNY leader*

***The content of the training program.*** New leaders need more than a group sharing last week's Workshop news. They need to rehearse a few pedagogical tools over the subsequent weeks of the term if they are to sustain their group's learning. Here is a menu and brief description of the teaching concepts we have found to be useful for new leaders. Probably no program can include all these components; directors need to shape their training plan to address the needs of their Workshop project and the resources available to them.

*The role of the group leader.* Your leaders have spent most of their educational years in the company of friends and under the guidance of teachers, and maybe tutors, teaching assistants, and advisers; the default is for leaders to model themselves on one these common roles. Although a good leader borrows from all these categories, this role is not identical with any of the above. Making these distinctions is important for the development of effective Workshop leaders. They are usually happy to discover that they are not expected to act like science professors, but they may need help in differentiating their roles from those of teaching assistants and tutors. The latter two categories tend to be more "top down" than that of a Workshop leader; the supposition is that TAs and tutors will fill in the weak spots in their students' knowledge base. Especially in the case of a tutor/student dyad, the assumption is that the student is in need of "help" and maybe even remediation.

It may be difficult for new Workshop leaders to embrace at first the larger goal of a good Workshop leader to guide students as they construct their own knowledge by making their way through the tasks set before them. It is an art form to be a nearly invisible guide. A perspective like the following can help them picture what kind of leader they want to be:

*As for the best leaders, the people do not notice their existence. The next best, the people honor and praise. The next, the people fear; and the next, the people hate. When the best leader's work is done, the people say, "We did it ourselves."* *Lao Tsu*

Leaders also need to think through the concept of boundary setting in several dimensions. In the first instance, they must consider how available they want to be for their students. On a residential campus, for example, Workshop leaders may find themselves under siege in the dorm by worried students the night before a big exam, and leaders on any campus can be buttonholed in the library or cafeteria by students who need reassurance about the latest problem set. Because they are usually amiable and receptive people, Workshop leaders can also end up on the receiving end of a great deal of personal information from troubled students. Therefore, leaders need to make some decisions ahead of time about the extent and limits of their duties. It also helps to develop a few effective deflecting strategies so that unhappy students can be routed toward appropriate sources of help when needed (more on this topic appears later).

Boundary setting also includes maintaining confidentiality. Certainly participation in Workshops is not such a private matter that it warrants the sort of protection accorded to visits to the doctor or to a counselor. But we also want to help leaders think through something like a golden rule for sharing information about their students. Just as we would feel very uncomfortable about having others idly discuss our performance in the hallways or over lunch at the student cafeteria, so too would our Workshop participants.

Another form of boundary setting entails the ownership of the Workshop participants' academic performance in the course itself. When students do poorly on exams, it is common for new leaders to feel responsible, and even guilty, for these low grades. They may feel that their own competence as leaders is gauged by their students' test scores. It takes some time for them to recognize that students as individuals are ultimately responsible for constructing their own understanding of the material and that a more genuine measure of effectiveness can be found in a leader's ability to keep students moving and engaged if they have encountered an academic setback.

Personal relationships between leaders and their group members are also part of boundary setting. A peer-led format means that, on occasion, leaders will find that close friends have been placed into their groups. Initially this change in roles may cause both parties a little confusion. We encourage our leaders in this situation to have a casual conversation with the friend in question, to clarify that their role is to be a coach, not someone who is judging students' performance in the course.

*The basics of group dynamics.* There exists an impressive bank of knowledge regarding the psychology of groups; the training director has much to choose from in this arena. Within this extensive domain, however, there are some concise and essential principles that can be delivered to your group leaders in an interactive, time-efficient way.

For example, new leaders need help in understanding how variations in personality and motivation have an impact on how groups function. Everyone knows that people are different, but leaders may not be sure how to transform these supposed obstacles into assets for the group as a whole. An easy way to engage your leaders in a discussion about differences in roles is to divide them into small groups and ask them to read brief scenarios like the ones following and describe how they would respond to students like these.

> Steve came into your group on the first day right on time and soon had his notebook and pencils at the ready. He was at work on the problems before you even started the session, so you were convinced you had a go-getter here. But, when it came time for the group to work together, he wouldn't speak up or contribute at all. Several times you glanced at his paper and saw he was on the right track, but every time you encouraged him to share something with the group or work with another student, he looked away or mumbled something about not knowing the answer
>
> Louise marched into the group the first day and announced to the group: "I don't know why we all have to be here. I already worked through every one of the problems the professor gave us." When you took a peek at her paper, though, you

could spot several places where she had the wrong idea. Nonetheless, she clearly seemed to feel sorry for the other group members and started to tell them how to do the problems. You were a little worried, first because she wasn't letting the other students do much of the talking, and second, because she was leading them down the wrong path.

Mara came into the room on the first day, sat down between two quiet students, introduced herself and said, "Wow, I just looked over these questions - they look pretty challenging to me! I'm so glad I don't have to face this stuff all on my own." After you started the group on the first problem, Mara turned to the person sitting on her left and said, "H'mm, what do you think? Am I on the right track here?" Soon the two of them were comparing their strategies and holding a friendly argument about the answer.

Lee came late on the first day of the Workshop and sat off to the side. He reluctantly joined the group in the center of the room when you encouraged him to move. He had forgotten his Workshop problems, so you had to give him your spare copy. The other students got right down to work, but you had to coax him into starting the first problem. When you asked for volunteers to come up to the board, Lee wouldn't even make eye contact with you. You hoped to touch base with him at the end of the session, but he was out the door before you had a chance to catch up with him.

Have each small group present their case study to the rest of the class and talk through the possible underpinning of behaviors like those described here. As the discussion proceeds, creating a $2 \times 2$ grid on the board can help leaders picture how group roles can be categorized:

| | **Initially positive** | **Initially negative** |
|---|---|---|
| **Outgoing** | *Mara* | *Louise* |
| **Quiet** | *Steve* | *Lee* |

During an exercise like this, several of your leaders are sure to say something like "I have a student just like this in my group!" If you have not done so already, this is a good moment to establish your rules of confidentiality for class discussion. In general, a good approach is to tell your leaders that during debriefing sessions they need to be careful not to mention the names or other personally identifying characteristics of their students.

These dissections of group dynamics help your leaders visualize how working relationships in a group evolve. These discussions also can help new leaders understand that how a group starts out is not necessarily how it will end up by the conclusion of the semester. *Group dynamics* means just that: a collection of people working together is constantly in the midst of change and development. Good leaders start to see how they can keep this change moving toward the mutual benefit of all the Workshop members.

*Student X and Student Y are the most conscientious - they had everything done correctly! I walked around and congratulated those that got answers early. Then I'd see someone who was stumped and would notice a similarity in thinking between two people. "Student A, you're so close - guess what? Student B made that same mistake!" Then they would work together.*
*Emily Lu, University of Rochester leader*

*Basic theory.* Our intention in teaching some of the rudiments of these theories is not to pack the training program with a great deal of abstract information. Instead, we want to select the most pertinent information about student development and then help our leaders make the best use of it during their Workshops.

For example, an especially useful theoretical model for Workshop leaders is Perry's (1968) description of the intellectual and personal development of college students. This theory, originally developed at Harvard in the 1950s-1960s, has provided a generative spark for several decades' worth of further investigation into the development of college students. In this model, students are viewed as travelers through series of progressive steps during their years in college. This series of positions traces a student's epistemological development from one in which authorities, including Workshop leaders, are seen to be in charge of the correct facts about their disciplines, whereas the students' responsibility is to place identical versions of these facts into the databank of their own memory system. Because students at this stage of intellectual development may have been especially attracted to the sciences as underclassmen because introductory courses often seem to reflect a way of knowing that is precise, organized, and well understood, they may not appreciate any weighing of options on the part of Workshop leaders; this will simply look like imprecision or a lack of knowledge.

But even those who enter college with an essentially dualistic perspective on the world are exposed to multiple points of view, both in and out of the college classroom. Such students are likely to be pushed to a point at which they begin to recognize that there are many perspectives about almost everything. This causes some distress, since they have to let go of the certainty of a rigid belief structure. In fact, at this point they can sink into an "anything goes" attitude about the search for knowledge. Soon enough, according to the Perry model, this blurry way of viewing the world becomes unsatisfying, and students start to recognize that, although everyone has a right to an opinion, not all these opinions are equal. And so the search for real understanding about the world around them begins. The theories, models, and constructs to which we devote so much of our time begin to seem valid or invalid not because the professor or the Workshop leader says that they are but because of the quality of evidence that exists to support these ideas.

The most compelling reason for bringing the Perry scheme into the training program is to help leaders understand that during the course of the term, students are probably doing more than just amassing science information and skills. If things are going right, they are also making a series of adjustments in they way the look at knowledge itself, and this is hard work. Perry made clear his admiration for the moral courage it takes for someone to construct a new point of view. We want our leaders to acquire the same patient respect for the development of their own Workshop students.

The research of Belenky et al. (1986) and Magolda (1992) provides a valuable extension of Perry's work. They surmised that Perry's scheme, developed from a sample of white, affluent males, might not map well onto the epistemological growth of a diverse female population. Belenky's team recognized that Perry's Harvard men often regarded themselves as future members of the power structure they saw before them (Authority-right-we). In contrast, many of their female subjects viewed authorities permanently as Others (Authority-right-them). This held true not just for the women from disadvantaged backgrounds but also for privileged women attending competitive colleges. They also found an epistemological position that did not emerge at all in Perry's study, a stage or level in which women are nearly completely without voices of their own. This way of knowing, which they term *Silence*, may be the most adaptive stance for some women, given their personal circumstances, but it is often not compatible with success in higher education.

Women who are able to emerge from this position may enter a time of subjective knowing; that is, they do not trust external authorities or the "facts" these outsiders have generated. Instead, they come to rely on their own intuition or instinct in making their way through the world. This approach marks a step beyond Silence, but it is still not one that meshes well with the intellectual demands of a college curriculum.

To help women continue their intellectual development, Belenky's team believes that good "connected teaching" offers the best link between the world of facts and ideas developed by external authorities and many women's ways of knowing. A connected classroom, in their view, poses problems that everyone, students and instructors alike, take part in discussing and solving together. It is one in which no one need apologize for being uncertain about the material, because uncertainty is understood to be an element of all human knowing. The connected classroom is one in which attempts at learning something new are not held up for challenge and competitive dispute but instead are encouraged and promoted ("What you're thinking is fine, but think more" p. 218). Sounds like a Workshop to us.

*I tried something new this week. I told the students to work on #2 with the group. Everybody was to contribute and the group was to come to a single answer. After they finished they were to come to the board and "present" their work. At first the room was quiet as students looked down at their own papers and remained to themselves. Then a voice mumbled here and there. Without warning there was an eruption of voices filled with discussion, questions, and critical thinking. I sat quietly and saw the group autonomously function. It was great!"*

*Gurjeet Birdee, University of Rochester leader*

Lev Vygotsky's insights into the relationship of instruction to mental development provide powerful ways of understanding the Workshop model. These and related ideas about social constructivism and cooperative learning are discussed by Cracolice and Trautmann in detail in Chapter 7.

Gagné's theory (1985) is also helpful to the Workshop leader because it emphasizes the hierarchy of subject matter. In this theory, founded on the information-processing model of cognition, the underlying prerequisite skills of a complex concept are analyzed, and then an instructional sequence is built from the simplest subordinate skill to the most complex idea. If the underlying skills are mastered, the entire concept should also be mastered; if an idea is misunderstood, an analysis of the subordinate skills can be conducted to identify the root of the miscomprehension.

Gagné proposes a sequence of nine instructional events that match the phases of learning according to information-processing theory:

- gaining the learners' attention
- informing the learners of the objective
- stimulating the recall of prior prerequisite learning
- presenting the stimulus
- providing learning guidance
- eliciting performance
- providing feedback
- assessing performance
- enhancing retention and transfer

Gagné's ideas are especially useful as tools for leaders as they prepare for upcoming Workshop sessions. As they think about addressing these steps with their students, they see a structure for the Workshop emerging.

Another very useful body of scholarship for the Workshop leader is the work done by Deci and Ryan (1990) on the nature of motivation. This work, described in some detail by Aaron Black in Chapter 8, "An Introduction to Theory and Research on Promoting Student Motivation and Autonomous Learning in College-Level Science," is particularly applicable to the Workshop. It characterizes well why this model is so effective in harnessing students' innate desire to learn and why the leader has a very special role in this process.

*Listening skills and questioning techniques.* Although a seasoned instructor may assume that successful students like our Workshop leaders already are capable listeners, we have found practice in listening skills a useful element for our training. In a small-group setting, listening to and understanding student questions is an essential part of the group dynamic. Learning to listen carefully is not a natural skill - it must be taught. Role-playing and case studies are simple ways to deliver this information.

Consider a scenario in which a student tells a leader that she doesn't understand a concept, say, net ionic equations. A typical student may express this as, "I just don't get this net ionic equation stuff." An untrained leader may take this statement at face value and begin a discourse on net ionic equations from beginning to end. This is certainly the wrong approach, however, as the true source of misunderstanding will almost always lie in some subset of the complete net ionic equation concept, perhaps some confusion about strong versus weak electrolytes.

Since any complete concept is described in the textbook and the course lectures, it is the leader's responsibility to extract the essence of the misunderstanding by carefully questioning the student while listening for a hint of the real problem. This is generally obtainable by working a specific example. A leader may respond to the student's generic question by asking, "Why don't we do a net ionic equation problem? You can then show me what it is that you don't understand."

Leaders need to be reminded that speaking speed is about one-fifth of the rate at which the brain can process speech. This can result in a wandering mind on the part of the leader and a resulting lack of understanding. To avoid this, leaders should be trained to *anticipate* student questions while they are being asked. This technique of guessing the question while it is being stated not only keeps the leader's mind engaged on the subject but it also allows him or her to come up with questions that the student may not be able to formulate.

Good listening skills involve more than just processing the words that are spoken; they also require being in tune with tone of voice and body language. Much of what we tell one another is not in words at all, so good leaders need to know how to attend to the nonverbal signals the students are providing, and they need to be aware of the cues they themselves are giving. The simplest role-playing exercises can demonstrate how a gap opens up in communication once eye contact is lost and a tone of voice seems disinterested.

As a counterpart to good listening skills, a few *questioning techniques* are also a must. It helps to talk over the negative impact of yes-no questions, and, conversely, the power of good questions. The leaders need to believe in the value of small steps; questions like the following ones can help Workshop students feel equipped to tackle the larger components of a problem.

- There are some unbelievably long words in this program. Let's go through it, just to make sure the pronunciation of all this stuff is clear.
- There are some symbols here that don't show up in everyday conversation! It would be a great idea to decode them all quickly - we don't want to walk out of here confused today.

- What is this question really asking us to do? What is the very first thing we need to do to get started on the problem?
- Give us your best guess - what is the most important information given in this problem?

Once leaders see how natural this sort of questioning can be, they are likely to feel relief: they really are not expected to be walking science encyclopedias.

*Learning styles.* A wise program director helps new leaders reflect on those nearly unspoken paradigms and models of instruction that they have absorbed without serious consideration or deliberation. Students selected to be Workshop leaders are by definition those who have excelled in tough introductory courses and have won at the game of science and math instruction, so, without intentionally establishing the re-creation of the self as a goal, they tend to work toward making clones of themselves. In many respects, this cloning may not be a disaster; the success Workshop leaders have had in the world of higher education validates to some extent the merit of their own approaches to the study of science. The tactics Workshop leaders have used to gain mastery over these subjects can be quite functional.

But this unconscious self-replication bears a cost: they fail to reach students who have different orientations to learning. Thanks to articles in the popular press (Gardner, 1993, 1994), there now exists something of a general understanding that we all may not acquire new information in the same way. A good way to launch this topic is to have the leaders complete a quick learning style inventory, perhaps Kolb's Learning Style Inventory II (1985), the Rogers Indicator of Multiple Intelligences (1995), or the Gregorc Style Delineator (1985). Leaders then can compare their results. They are bound to discover that, even in a group with similar academic interests and level of success, there is broad variation in preferred learning modes.

As a follow-up, leaders can be asked to craft a Workshop exercise that teaches the same concept in several different ways to match different learning styles. For instance, a problem about kinetics could ask students to focus on quantitative approaches to learning this concept; a variations could be one that involves a simulation, and another that requires students to write out verbal responses. Ultimately, of course, we hope that our students master concepts from several different angles, but leaders should be able to help all types of learners over the first hurdle of learning new materials.

*Reflective teaching***.** We want our Workshop leaders to become professionals who are able to consider and reconsider beliefs about teaching, learning, and their subject matter. They need to be able to make thoughtful decisions about how they lead their Workshops and to develop strategies for making good decisions about what they will encounter with their groups. Good leadership requires continuous reflection on the progress of coaching the groups and developing leadership skills, but thoughtful contemplation is hard to come by in the midst of our leaders' packed academic and personal schedules. Our responses to their journal entries are important in helping our leaders consider the current health of their Workshop and to take any necessary steps toward improving the functioning of the group. We can help leaders by constructing assignments and exercises in our training programs to assist them in their growth as "reflective practitioners." One such assignment is the writing of an autobiographical essay. When writing their autobiography, Workshop leaders may start by considering the following prompts:

- What stands out for you about your early education?
- What are some recent striking memories related to your academic life?
- What has been easy for you to learn? Why has this gone so well?
- What has been more difficult for you to learn? What has made this hard?

- Do you think your educational experiences might have been different if you were of a different ethnicity or gender?
- In your view, what are the qualities of an educated person?
- How well has college helped you become an educated person so far?

Another section relates more directly to the Workshop experience. If assigned at the beginning of the term:

- Why do you think you were chosen to become a Workshop leader?
- What are some of your worries about being a leader?
- What are you most looking forward to about this experience?
- How do you expect to help your Workshop participants?
- What might your students teach you?

If assigned later in the semester:

- What has been the biggest surprise about being a leader?
- What has been the most important way you've helped your students?
- What have your students taught you?

Another activity that provides leaders an opportunity for reflection is to have them lead a brief reflective teaching lesson to a small group of their peers within the training program. This is similar to the practice Workshops described previously in the section "First Things First." The tone and content of this activity are different, however, once leaders have had the opportunity to develop a little expertise in running a group.

A lesson of this sort helps leaders engage in the complete act of facilitating a group: planning, coaching, and evaluating. During the reflective session, one or more of the students are selected to be the leaders for the day, and small groups of the others are the learners. Each designated leader then conducts a session, using a few of the Workshop problems from that week.

When the session is completed, each group takes about fifteen minutes to reflect on the coaching and learning process. Discussion may focus on the *planning process*; that is, the knowledge, skills, and attitudes they were hoping to develop in their students, and the resources they used as they prepared for this session. Encourage them to describe what they would change if they were to redo that session.

The *learning process* should also be part of the discussion; the "learners" (the leaders serving as students) can reflect on what the leader did to help them. This feedback should include information about how comfortable they all felt about answering and raising questions and trying their hand at Workshop problems.

The desirable outcome of an activity like this, of course, is not only to advance the development of the leaders who are in the "hot seat" for that day but also to promote the growth of leadership skills for the whole group of newcomers. For more on reflective teaching practices, see Schon (1983, 1987) and Dewey (1933).

*Teaching tools.* From the vast array of pedagogical strategies, program directors will find particular tools to be good matches for the format and mission of a Workshop program. Our leaders and Workshop students have responded especially well to the following techniques. Many of these tactics are described in more detail in Chapter 3 and in Appendix I.

*a. Concept mapping.* Concept maps, written diagrams that list a number of words or phrases describing a group of related ideas, are powerful teaching and learning tools. As you can see from the example of a concept map in Chapter 3, "Writing the Workshop Materials," the meaning of concept maps is represented by connecting lines between the concept statements and the brief statements of the linking relationships. Please look at this concrete example before continuing your reading.

Concept maps help students see the relationships among a group of ideas. Research in educational psychology shows that the better the mental connections between concepts, the more likely these concepts can be recalled and applied. Constructing a concept map can be an excellent cooperative learning exercise because it gives students an opportunity to challenge and expand the mental networks that they and other members of their groups hold.

Your leaders can be taught how to coach their students through the development of a concept map early in the term for a theme that runs throughout the course. Your group's initial concept map, perhaps on stoichiometry in first-semester general chemistry, could be constructed at the Workshop immediately following the conclusion of the stoichiometry lectures. As the term progresses and new stoichiometry concepts are introduced, leaders can bring out their groups' concept maps and have them make modifications based on their new understanding. In a typical first-semester general chemistry course, this would include aqueous solution stoichiometry, gas stoichiometry, and thermochemical stoichiometry. Such a concept map has the potential to improve greatly the students' understanding of the relative simplicity of the stoichiometry pattern, which otherwise may appear as a group of independent concepts.

For more on this topic, see Cliburn (1990) and Regis and Albertazzi (1996).

*In a Workshop, I did a concept map with my group and found that students came up with different concept maps of the same materials. Astonishingly enough I also found that those concept maps gave them a chance to examine their understanding of the materials.*

*Shah Marui, CCNY leader*

*b. Model building.* Since we cannot directly sense atoms and molecules, we use models to provide a way of creating a mental image of these tiny particles. Once we have constructed a model, we can use it to make predictions. Models either can be purely conceptual, such as in the formation of a mental image of the quantum mechanical model of the atom, or they can be partly conceptual and partly physical, as in the molecular model kits, which are physical representations of mental constructs of molecules, and are commonly used by general and organic chemistry students.

Model building is important because it is the essence of the scientific method. A scientific inquiry typically begins with the collection of data about the natural world. These data are then used to build a model: a mental construct, or hypothesis, explaining the data. The model is then used to make predictions. If situation X is true, then Y must result. These predictions are then subjected to tests. If the tests support the validity of the model, the model gains strength. If the tests refute the model, the model must be modified or discarded.

In order for your students to absorb the philosophy underlying science, you must emphasize, whenever possible, the role that model building holds in science. The *if-then-therefore* pattern found in scientific thinking is the type of cognitive skill we want our students to develop in our courses.

c. *Pair problem solving* (Herron 1996; Whimbey and Lochhead 1986). In this method, the leader divides students into pairs and designates one person to serve as the problem solver and the other as the "checker." The problem solver reads the problem and continues by talking through his/her thoughts on the way to the solution. The checker is there not as a critic but as a guide and as a listener.

The purpose of this approach is to make thinking visible and to ensure that ideas are detailed and reviewed at each step along the way. This careful elaboration and listening is not an approach that is immediately understood by many students, however, so leaders need some practice with it before they bring this strategy to their own groups.

*Their own research.* During the course of the training program, and most especially as they lead the Workshop sessions themselves, new leaders come up against all sorts of interesting problems and issues that they feel are not fully addressed within the training format. These questions and dissatisfactions offer us an excellent opportunity to ask our leaders to pursue these interests, with the goal of sharing their newly acquired information with the rest of the group later in oral presentations, poster sessions, and written essays. If the training program has been formatted as a credit-bearing course, it is easy to include a project of this sort as one of the course objectives.

We have found these individual projects to be of real value to the program. For instance, some leaders have designed microteaching units for their Workshops about especially tricky concepts, like stereochemistry, or crossover in genetics. When this information is shared in the training course, leaders take back to their own Workshops the material and strategies that they have learned from their colleagues, reinforcing at another level the idea of student-led learning.

Other types of projects investigate a medium or format that supports learning, for instance, reviewing a set of computer-aided instructional programs and appending a list of recommendations about purchases for the learning center. On a different tack, leaders often elect to take a closer look at the impact of demographics by examining the influence of race, gender, and culture on Workshop participation. These projects often include the input of the Workshop members themselves through questionnaires, focus groups, and individual interviews. As project directors we have learned a great deal about the model by examining the information our leaders have compiled for this assignment.

The student projects also promote the professional development of the leaders themselves. As most of them are enrolled in science curricula, they frequently have had limited opportunity beyond their freshman English class to complete research of *this* sort. The project gives leaders a structured incentive for using the latest information retrieval systems and conducting their own personal investigations. It also offers a venue for presenting this information to their peers through written reports, oral presentations, and poster sessions. If the training program is a team-taught enterprise, feedback can be given on both the content, from the science instructor, and on methodology, from the learning specialist - all good practice for their future professional responsibilities.

*There are a number of advantages or benefits that we earn from cooperative learning: friendly relationships among different ethnic groups, respecting and supporting in Workshop groups, motivation of peers - moreover, a feeling of attachment.* *Francis Kim, CCNY leader*

*Race, class, and gender issues.* We would like to believe that cultural and social tensions will be checked at the door of our Workshop rooms so that all students can work together on equal footing for their common good. Of course, this is often not the case. Conflicts related to race, class, and gender permeate our educational environments, just like they pervade the world beyond the walls of academe. We should not expect our leaders to negotiate this uncertain social territory without a little guidance from us; creating an awareness of gender issues and cultural differences belongs in a leader training program.

To start, try the following simple exercise with your Workshop leaders. Ask them to draw a picture of a scientist. Some students may be inhibited about their artistic skills, but encourage them to do their best anyway. Tell them to put in all the details of dress and hairstyle they can.

Unless students have been through this exercise before, the results are all too predictable: most of them will invariably draw a male scientist, often with an Einstein hairdo, glasses, and a lab coat. Very seldom do students draw a female scientist or one of color. Nor do they tend to draw themselves. Even though we are often working with science majors, they are compelled, like the rest of society, by the stereotypes that are presented through television, the movies, and cartoons.

When this is drawn to their attention at the conclusion of this exercise, most students are a little surprised at themselves. They tend to remonstrate with us at some length, saying that they did not really mean to portray things this way. This affords us a powerful teaching opportunity, for at that moment we have an opening to talk about all those other unconscious, inadvertent things we do that pigeonhole other people by virtue of demographic category.

Another approach is to present a few choice quotations, like the following ones. Comments like these usually fuel considerable, and sometime painful, discussion. Be prepared to help them stay on track as they talk these things through.

> *"There is something about problem solving or the required abstract thinking that somehow deters women from pursuing a career within the field of science."*
> (from a female student)
>
> *"I'm in a group where the ladies are very aggressive. The guys are aggressive too, but some of the ladies are quicker to call out the answers. This is typical in the 90s girls' attitudes. They often feel that they can be equal to or sometimes better than the males."* (from a male student)
>
> *"I don't think race is a factor in my group in any way at all."*
> (from a majority culture student)
>
> *"We shouldn't be so quick to alter the way science has been handled in our culture. After all, the old way of doing things gave us modern medicine, Teflon, computers, and so on and so on. Why should we change what isn't broken?"*
> (from a male scientist)

The larger goal here, of course, is to help the leaders create an open and encouraging learning environment in the Workshop for all students. This requires that they attend to their own cultural backgrounds and their comfort level with students from both genders and different cultural groups. It also demands that the leader attend to these dynamics among the Workshop participants and be prepared to respond quickly and effectively if these differences become tensions within the group.

Our campuses often have resources to help us become more aware and skilled in these issues; check with your office of minority student affairs and your Equal Opportunity officer. The staff members in these programs may be able to supply you with print resources, and they may be willing to make a presentation at one of your meetings with the leaders. They are also good referral sources if matters become truly conflicted and some sort of mediation is in order.

*Disability issues.* At this point in American educational history, our institutions and programs often make claims that they welcome a diverse range of students. Most institutions make statements like, "Big Pine University offers equal opportunity regardless of gender, age, natural origin, color, creed, sexual orientation, or disability." But turning these abstract mission statements into genuinely accessible programs for our students is all in the details. For instance, making it possible for students with hearing or visual impairments to communicate with their Workshop teams often takes planning on the part of the program director, the group leader, perhaps the disability support office, and certainly these students. Similarly, leaders need strategies for helping students with dyslexia become full members of their Workshop sessions.

It is estimated that 6%-7% of our current postsecondary students have a disability (National Science Foundation, 1996), so in a general chemistry course with an enrollment of 100 students, about a half dozen students are likely to have a disability covered by the Americans with Disabilities Act (ADA). Although most of these disabilities are mild in nature, and probably require only modest accommodations in a Workshop setting, the number of college students with substantial disabilities is steadily rising.

We can expect some of these students to become part of our Workshop programs, but we cannot assume that our leaders will spontaneously understand the nature of these disabilities and the appropriate accommodations when the need arises. In fact, we cannot assume that the students with disabilities themselves - especially those in first-year courses - will have a complete grasp of their situations.

Fortunately, the legal imperatives of the ADA have encouraged most of our administrators to make sure that there is some expertise on our campuses in these matters. Check with your office of disability support, Equal Opportunity program, or learning center to see if a staff member is available to assist you and your leaders. These experts often can make presentations to your group of leaders and help out with some creative problem solving related to specific needs. Consult the many print resources, like *Working Chemists with Disabilities* (Blumenkopf et al. 1996) for more insight.

**Assessing the Training Program**

Many of us quail a little at the thought of evaluating a program of this sort; it seems difficult to come up with hard and fast measures of effectiveness when we are dealing with so many human-centered variables. But a simple chain of reasoning can help us collect the kind of information we need to sustain and improve our training programs. We need to begin with asking ourselves *why* we want to measure any particular component of the program. Clearly, the nature of the proof needed is different if we are justifying expenditures to our administrators versus helping our leaders make midcourse corrections in the way they are facilitating their groups.

Let us review two different assessment approaches, one summative, the other formative. Summative assessments might be applied in order to prove the worth of the training program to our deans and provosts, as well as to ourselves. The goal here might be to make a case for continued funding or to demonstrate the value in expanding the program. In conducting research of this sort, your campus office of institutional research or your registrar should be able to help. Merging the assessment of the Workshops themselves and the training program is important in verifying cost effectiveness. If the registrar's information shows a small but statistically significant improvement in the Workshop participants' performance and retention in the course, then it is easy to demonstrate the cost effectiveness of the Workshop and the training; that is, it is expensive to replace students; it is cheaper to build the kind of support that keeps them in our classrooms and matriculated in our degree programs. An examination of your campus database can help you establish a profile of your Workshop leaders. We want our administrators to know what a remarkable group our student leaders are. Establishing the leaders' value to the institution in turn validates the feedback we collect from them through focus groups and questionnaires. As our leaders are some of the most sophisticated students we have on our campuses, their opinions are worth a great deal.

Perhaps nearer to our own hearts is the kind of information we derive from *formative* assessments. In these cases, our goal is to make midterm improvements in the quality of the training program, while it still counts for that particular group of leaders. So we want to look at their performance in the Workshops and their own satisfaction rates with the training they are receiving. We can get at this information first by trusting our leaders to tell us what they need. Again, the leaders' journal entries about their Workshop sessions are probably our best ongoing mechanism for gathering information about their skills. They tell us how they solved problems, or failed to, and the level of their insight into group dynamics is often made very clear through their own words.

Additionally, we need a direct window into the Workshop sessions themselves. It is probably intimidating for our new leaders if the program directors sit in on the Workshop sessions. In fact, our presence can easily undermine the nature of peer leadership - which, after all, is the essence of the Workshop model. A better strategy is to have our Workshop leaders visit one another's groups, fill out a brief questionnaire about their observations, and then have the observer and the observed spend some time together discussing this experience. (An observation assignment is included in Appendix II.) Our super leaders can also fulfill this role; since they are students too, their presence in the Workshops is less stressful for the leader and less disruptive to the group.

Part of formative assessment needs to include the Workshop participants' opinions about the effectiveness of their groups, which are most easily gathered through questionnaires. We need to remember that however accustomed we may be to reading student opinion surveys, this is likely to be our leaders' first occasion to be at the receiving end of this sort of data. They may need a little guidance to avoid feeling defensive and to make good use of this feedback.

Formative assessment, of course, should continue through to the conclusion of the term, as our goal includes the improvement of the training program for future semesters as well as the current one. End-of-term questionnaires are important for both the leaders and the Workshop participants themselves. For more about the assessment of Workshops, please see Chapter 6 in this *Guidebook* "Workshop Chemistry Evaluation."

*I'm always happy to be in a group, working together, having ideas about different kinds of people who are living in the United States. As a Workshop leader I see my role as a teacher and a learner, a leader who will be responsible to lead the students in the right path, a friend of the students who will help [other] students in time.* *Shah Maruf, CCNY leader*

## References

Belenky, M. F., B. M. Clinchy, N. R. Goldberger, and J. M. Tarule (1986). *Woman's Ways of Knowing: The Development of Self, Voice, and Mind.* New York: Basic.

Blumenkopf, T. A., V. Stern, A. B. Swanson, and H. D. Wohlers, Eds. (1996). *Working Chemists with Disabilities: Expanding Opportunities in Science.* Washington, D.C.: American Chemical Society.

Cliburn, J. W. (1990). Expanding Maps to Promote Meaningful Learning. *Journal of College Science Teaching, 19*: 212-215.

Deci, E. L. and R. M. Ryan (1990). *A Motivational Approach to Self: Integration in Personality.* In *Nebraska Symposium on Motivation.* Lincoln: University of Nebraska Press.

Dewey, J. (1933). *How We Think.* Boston: D.C. Heath.

Freire, P. (1971). *Pedagogy of the Oppressed.* New York: Seaview.

Gagné, R. M. (1985). *The Conditions of Learning and Theory of Instruction.* Fort Worth: Holt, Rinehart and Winston.

Gardner, H. (1993). *Frames of Mind: The Theory of Multiple Intelligences.* New York: Basic.

Gardner, H. (1994). *Multiple Intelligences: The Theory in Practice.* New York: Basic.

Gregorc, A. F. (1985). *Inside Styles: Beyond the Basics.* Columbia, Conn.: Gregorc Associates.

Herron, J. (1996). *The Chemistry Classroom: Formulas for Successful Teaching.* Washington, D.C.: American Chemical Society.

Kolb, D. A. (1985). *Learning Style Inventory: Technical Manual.* Boston: McBer.

Magolda, M. B. (1992). *Knowing and Reasoning in College: Gender-Related Patterns in Students' Intellectual Development.* San Francisco: Jossey-Bass.

National Science Foundation (1996). *Women, Minorities, and Persons with Disabilities in Science and Engineering.* Arlington: NSF.

Perry, W. G. (1968). *Forms of Intellectual and Ethical Development in the College Years: A Scheme.* New York: Holt.

Regis, A. and P. G. Albertazzi (1996). Concept Maps in Chemistry Education. *Journal of Chemical Education 73*: 1084-1088.

Rogers, J. K. (1995). *Rogers Indicator of Multiple Intelligence and Scoring Grid.* In Corey, G., Corey, C. and H. J. Corey, *Living and Learning*, pp. 36-40. Belmont: Wadsworth.

Schon, D. (1983). *The Reflective Practitioner: How Professionals Think in Action.* New York: Basic.

Schon, D. (1987). *Educating the Reflective Practitioner.* San Francisco: Jossey-Bass.

Whimbey, A. and J. Lochhead (1986). *Problem Solving and Comprehension*, 4th ed. Hillsdale, N.J.: Erlbaum.

# *Chapter Five*
# *Institutionalizing the Workshops*

**Jack A. Kampmeier, The University of Rochester**
**Pratibha Varma-Nelson, St. Xavier University**

Although students seem to change their plans and their majors about as often as they change their socks, the faculty and the institutions that encourage those changes often seem to be immutable. We continue to organize by disciplines, reward by disciplines and teach in much the same way we have always taught. This chapter is about change, in faculty and institutions. It is about converting effective practices, often supported with some short-term funding, into continuing practices.

## Identify the Problem(s)

If faculty are satisfied with the mechanisms and the outcomes of their teaching, neither evidence nor rhetoric will change their minds or their habits. In fact, the situation is even worse than that; any discussion that starts with the idea that "Workshops are better" will elicit an incredible barrage of explanations, justifications, rationalizations and denials, all designed to defend against the perceived implication that "If mine is better, then yours is worse." Many discussions about change begin and end at this stage, for good reason. Most faculty are trying to do a good job in the classroom.

On the other hand, most faculty are also trying to do a better job in the classroom. Because the best impetus for change is a nagging problem, the challenge is to identify the areas of faculty dissatisfaction and to connect the Workshop model to those concerns. The goal is to get faculty thinking that Workshops could be a way to solve the problem that is bothering them. The specific problems will be as varied as the strengths of the Workshop model. The classic complaint is that "They just don't get it." Maybe that is because the structure of the situation invites passivity rather than activity. Or, "They seem to hate it." Maybe it would help to have peers working as cheerleaders for the subject and the course. Or, "They don't know how to study." Maybe we need to find mechanisms to help students learn how to study. Workshops can help with all these problems.

In addition, different courses on different campuses present different problems. For example, first-semester students at an urban commuter campus may need help in making connections to one another and to the campus. Third-semester organic students at a residential campus will not have that problem but will have other problems. In general, organic chemistry presents a set of challenges that are quite different from those in first-year chemistry. Nevertheless, the Workshop model is very robust and successful in a wide variety of situations. Regardless of the specific context, students have to construct their own understanding, and the Workshop model facilitates that process.

For introductory science courses, the issue of retention cuts across administration and faculty. The problem is probably clearest to administrators. Although the nature of the response varies with the nature of the school, the institutional consequences of student failure or dropout are generally negative. Income is lost, public support is weakened, time and effort are wasted, and human potential is unrealized. This is neither good for the institution nor the faculty. In the end, faculty livelihood depends on support from families, taxpayers, and donors. This support is provided in expectation of positive results.

To be sure, student success has much more to do with personal and intellectual development than it has to do with retention. It is the responsibility of the institution to stimulate and support that development. There is something wrong with failure and dropout rates that approach 40%, just as there is something wrong when successful students turn their backs on subsequent courses in the discipline. There is something wrong with an educational institution that functions as a disinterested gatekeeper and approves only those who manage to jump a certain height. It is the job of the faculty to provide structures to support the development of the personal and intellectual skills required to jump that certain height. Workshops can provide that support.

In times of decreasing resources, universities must find ways to do more with less. Real productivity gains come from improvements in quality. When students are more successful and more satisfied with their experiences, in a context of constant or increasing expectations, then the quality of education is improved. and the faculty and the institution are more productive. Workshops do just that. The productivity gain has many sources. The most important new source comes from the students themselves. By harnessing the power of debate and discussion among peers, the Workshops promote the construction of individual understanding.

There are many problems that concern faculty and institutions. Workshops will not help with all of them. They will, however, help with many. The road to permanency is to identify the connections between the Workshop model and the problems that are nagging the faculty and the administration. The purpose of this first stage of the discussion is to identify the nagging problems that might be treated by Workshops.

**Explain the Model**

In our experience, it is essential to get enough time to state the case for Workshops. At a minimum, one needs to explain what a Workshop is and what a Workshop is not, and to provide evidence that Workshops work. The problem is that the casual listeners almost always jump to the conclusion that the Workshop is something they have been doing for years or something that they already know about, dressed up in new clothes. Unless the model is sharply and aggressively defined and described, it gets equated with tutorials, small recitation sections, process learning, cooperative learning, supplemental instruction, discovery groups, or some other structure with which the listener is already acquainted. Although the Workshop may share same elements with some of these other approaches, the model has a distinctive structure and purpose, and a distinctive theory and practice.

A very common misunderstanding is to equate Workshop leaders with teaching assistants (TAs). Since Workshop leaders are usually undergraduate students, the leader is mistakenly conceptualized as an undergraduate TA or a mini-TA. This leads to trouble in short order. We usually describe our graduate TAs as knowledgeable young professionals who help the faculty explain the subject to beginning students. We, and others, expect TAs to be well on the way to mastery, to know their subject matter and even to have some skills in presenting the material to others. An undergraduate Workshop leader who just took the course last year cannot possibly meet these expectations. It would be quite proper to object to the use of undergraduates in this teaching assistant role. With sufficient time, one can explain that the Workshop leaders are not just inexperienced TAs. The undergraduate leaders are expert about learning and this is the quality that enables them to be legitimate guides, facilitators, cheerleaders, and mentors. The leaders do not work as graders or answer givers, because they do not have the authority that comes from being expert.

Others will think that the Workshop is a place to go to get the answers to the homework, like a recitation section or a HELP room. Although the Workshops are usually built around problems, the focus is not on answers. Rather, the Workshops are designed to engage the students with the subject and with one another. The point is not to be shown how to do it but to grapple with finding out how to do it by actively working on the problem in a context that encourages, supports and even requires questions, discussion, and debate among peers. In a well-functioning Workshop, the leader might not seem to do much more than get the group started, keep track of the time, and orchestrate a wrap-up that gets the group discussing the take-home points. In fact, the leader is probably also operating throughout to encourage and support each member of the group, to monitor their participation and contributions, and to intervene in subtle ways to make sure that everyone is catching on. In a really good Workshop, it is hard to tell who is the leader. If the leader stands up and starts to "teach," the Workshop is probably broken.

In summary, the purpose of this stage of the process is to bring the listener to the realization that the Workshop model is something new and different in structure and purpose. It is not just the same old stuff in a different package.

### Connect the Workshop Model to the Problems

Once the nagging problems are identified and the Workshop model is explained, the next step is to link the two. This is the time to explain the model more fully, to provide theoretical and practical reasons for believing it will work, and to show just how *Workshops* will help with the *problems* at hand. This is the time to roll out previous results. The purpose of this stage of the process is to convince the listeners that Workshops are in *their* best interests. The process is modeled in the following paragraphs.

Beginning students face a variety of personal and intellectual challenges. Because they are not yet integrated into a community of learners, they often face these challenges alone. Workshops can provide a ready-made learning community and a model of what it means to be a working member of the group. At the same time, the Workshops model the way we think learning occurs. Students often come to us with a gas-station model of education: college is the place to go to get filled up. The lecture seems to reinforce that passive notion. The Workshop model is designed to emphasize that learning is an active process in which students program and reprogram their minds. This is a creative process, built on question, debate, discussion, and interaction.

It is clear that different students learn in different ways. As a result, it is important to provide students with many different kinds of opportunities to learn. The Workshop does just that; students can build models, organize tables of data, make plots, calculate, draw pictures, argue with one another, present at the board, build consensus, compete as teams, explain to one another, ask questions to one another, work cooperatively, complain, and celebrate. A skilled Workshop leader encourages a range of activities, recognizing that there are many different paths to understanding. Above all, students want to express their individuality, even if it is just their own puzzlement. The best Workshop leaders know this and work hard to encourage and support that drive for autonomy.

Beginning students need mentors; the Workshop leaders naturally fill those roles. The leaders were chosen because they are very successful college students, with established records of learning in a variety of situations. They were also chosen because they have the personal qualities that fit the role of mentor and trusted counselor. Simultaneously, the leaders function as role models for younger students. They do what most students want to do: they do very good work and they have fun.

Today's employers continue to want employees who have specific disciplinary skills. However, those skills are necessary, but not sufficient. In addition, employers place high value on communication skills and the ability to work in teams to solve significant problems. The Workshop model is designed to facilitate the development of just these skills. The Workshop emphasis is not on *the answer* but on how to get the answers, how to evaluate answers and how to recognize and choose the best answers. In the process, students are required to explain, justify, negotiate, construct, modify, listen, and learn. These are transferable, marketable skills.

Colleges and universities need mechanisms for acknowledging the success of their best students. They also need to provide opportunities for new growth to those students. College is not just piling up bricks, four per semester. Each semester must present different challenges, in graded, purposeful ways. The opportunity to become a Workshop leader is both reward and opportunity. The leaders are invited into new relationships with their peers, the faculty and the college. They are invited to become responsible for making things work. This is a big event, transforming in its implications and consequences. It has the potential to change the way that students construct their own education and the potential to change values and goals. If for no reason other than to provide reward and stimulus to the best students, good educational settings need Workshops.

**Assess and Evaluate**

Your enthusiasm for Workshops will certainly be challenged by your colleagues and by your administrators. They will want "convincing evidence" before they will commit their time and energy or institutional support to your project. In order to respond to their questions, you will need to assess the effects of your own work. Two preliminary notes of caution are important: (1) A classroom teaching experiment is not like a scientific laboratory experiment. Ideas about precision, accuracy, and controls are quite different; (b) You are an expert about science and teaching, but you are probably not an expert about educational research. Do these related notes of caution mean that we cannot evaluate the effects of Workshops? Certainly not. But, they do warn us to be forthright about the limitations of our methods and our results and not fall into the trap of trying to meet laboratory standards in a social experiment.

Assessment and evaluation have been part of the Workshop Project from the start. Our methods and results are described in detail in Chapter 6. Questionnaires and survey instruments are provided. In general, there are two kinds of tactics that are easy to use and provide good insights. One involves asking the students what they think. The other involves counting.

The opinions of students are important and, by definition, valid. They are, after all, in the business of taking courses. Their opinions do not reveal whether a course *is* a "good" course; they do reveal whether the students *think* it is a good course. Students have experience with a variety of teaching methods. It makes sense to ask them what they think of the Workshop method. The Workshop leaders form a special student group. They are good students and sophisticated observers of our teaching methods. Their opinions carry special weight.

Standardized college questionnaires are useful because they provide reference points in other courses and norms for the college. Standardized Workshop questionnaires can provide opinions about a specific Workshop course that can be compared with opinions about other Workshop courses in the larger Workshop Project (see Chapter 6). Student opinion questionnaires usually focus on matters about delivery and structure. The cost/benefit section of this chapter emphasizes the need to assess the amount of learning that occurs in different teaching situations (for example, Workshop and non-Workshop). Elaine Seymour has developed a questionnaire that focuses on the students' sense of their learning gains (Seymour,1998).

We have also used other techniques to gauge the effects of the Workshops on student learning. The simplest of these counts the percentage of students who *successfully* negotiated the course with a C$^-$ or better.

$$\text{Percent Success} = \frac{\text{number of A, B and C grades}}{\text{total number of students}} \times 100$$

The "total number of students" is defined as the number enrolled after the no-penalty drop period; it includes all those who are ultimately assigned letter grades (A, B, C, D, E) plus those assigned Incompletes and those who dropped after the no-penalty date. The trick in using Percent Success to evaluate the impact of Workshops comes in the choice of reference. In the best case, the same faculty member has taught the course for many years and the Percent Success *before* and *after* Workshops can be compared. In other cases, there may be multiple sections of a given course, with different teachers but common exams. Percent Success can then be compared in Workshop and non-Workshop sections of the course. Such data must be used with great sensitivity because they are dangerously close to comparing one teacher with another. In yet another setting, there may be experimental Workshop sections embedded in a larger non-Workshop course with a single teacher. If care is taken to keep grading standards constant and to watch out for big differences in the makeup of Workshop and non-Workshop sections, comparisons of Percent Success can give an important measure of the impact of Workshops.

If the grading standards remain constant, the average grade earned in Workshop and non-Workshop settings is another measure of the impact of the Workshops on learning. With proper respect for the caveat that grading standards remain constant, this measure can be applied in the different settings outlined above.

Finally, we have used attendance data to measure student opinions about Workshops. Students are very busy and jealously guard their time; in general, they do only those things that they believe to be directly related to their immediate needs and goals. When they are not artificially constrained, students vote with their feet. Attendance data for Workshops is quite astonishing (>80%) and provides sturdy, utilitarian evidence for their value.

**Your Job Is to Teach Well; The Institution's Job Is to Reward You**

There are three elements to the work of a faculty member: teaching, research. and service. Although institutions of higher education value all three, discussions about promotion and tenure often focus heavily on research. As a result, junior faculty members often wonder if their efforts to integrate Workshops into a course will be rewarded by promotion and tenure. The work to develop and implement a Workshop program can strengthen teaching and learning, provide a documented record of creative research and offer opportunities to serve. It is in the faculty member's interest to insist that these contributions are recognized and rewarded.

While it is the responsibility of faculty to utilize methods that produce effective learning, "the fastest and most enduring way to promote a renewed emphasis on teaching in the service of learning is to restructure the faculty rewards system" (Seymour 1998). There is an extensive literature that supports this view and a vigorous discussion of methods to bring about the necessary changes (Moore 1997; Coppola 1996; Fiesel 1995; Boyer 1990; Glassick, Huber, and Maeroff 1997; Guskin 1994; Schon 1995; Boyer Commission report 1998; Paulsen and Feldman 1995). A broader definition of teaching that is not restricted to lectures must be adopted by those who make decisions about promotion and tenure. Boyer's expanded view of scholarship should be adopted to recognize and reward the effort involved to develop and implement innovations in teaching. Evaluations need to be structured to reflect a variety of scholarly activities and teaching methods.

Barr and Tagg (1995) argued that if colleges and universities are to change from institutions that provide instruction to institutions that produce learning, then the faculty member will have to change from being an actor - a "sage on stage" - to an "inter-actor" - a catalyst that promotes learning through a variety of interactions among the students, the teacher, and other role models. Implementation of Workshops is a way of making the transition from actor to inter-actor.

If you introduce peer-led team learning, you need to be prepared to facilitate your own teaching evaluation. Clearly report your role in the Workshop model of teaching. Document your time and effort. Make sure that your administrators and colleagues are aware that:

- You are collaborating with peer-leaders to create a student-centered Workshop environment that addresses a variety of learning styles.
- You are involved in selecting and training the peer leaders in order to keep the Workshop sessions focused and productive. Explain whether you are the only one involved or if you do this in collaboration with a learning specialist. Report selection procedures and include training materials that you develop in your portfolio.
- You are developing some of your own Workshop units. Include samples in your portfolio.
- You design, coordinate, and supervise all aspects of the course. Document the amount of time spent on each of these activities in addition to the time spent on preparation and delivery of lectures. Although the faculty member is not physically present in the classroom during the Workshops, the effort that goes into the effective operation of the weekly Workshops must be included in calculation of workloads. (This can be especially important at liberal arts colleges and community colleges where a teaching load is defined in terms of the number of hours spent in the laboratory or classroom.)

At most institutions the end-of-semester questionnaires are identical for all courses, with little consideration given to differences among disciplines or types of courses (Shulman 1993). This may be appropriate when the method of instruction is the same in all courses; that is, faculty lecture and students listen. Since the Workshop is fundamentally different, in structure, purpose, and practice, different kinds of assessment instruments are needed. Most student opinion surveys evaluate delivery. A Workshop evaluation might ask how well the faculty member has designed and managed the learning environment and whether the Workshop program promotes discussion and learning. If you are required to use the college evaluation form, perhaps you should do two evaluations simultaneously: one to satisfy the institutional requirements and the other to evaluate the effectiveness of Workshops. Make sure that the reviewers of student opinions realize that you are not present in the Workshops and hence are less visible to the students. This is especially important at institutions without graduate students, where the faculty traditionally conduct recitations. The reviewers also need to recognize that students can be uncomfortable in their modified roles as active participants in Workshops.

For teaching evaluations to be meaningful, they must take on a variety of forms. Classroom visits by colleagues and administrators are helpful complements to student questionnaires. It is important for evaluators to visit both the lecture and the Workshops. One visit may not be adequate, so careful observation over a period of time may be necessary. A clear understanding between the visitors and you must exist on such matters as how the Workshop functions, the role of the peer leaders, how the Workshop materials are designed and selected, what you are trying to accomplish with a particular unit, and the training you provide the peer leaders. The observer must be aware that the atmosphere that exits in Workshops is very different from that in lecture. Peer leaders should be informed of the visits ahead of time.

Cooperation among peer leaders, students, you, and the administration is necessary to obtain reliable data.

The Workshop Project can be a framework for research in science education. The PLTL model is particularly suited to these purposes, since it has a sound theoretical base. As all good research should, the work to date has opened up possibilities for further exploration and expansion. There are many opportunities to test the model by providing data about Workshop students and Workshop leaders. There is always need to measure learning gains. Presentations of conference papers, seminars, peer-reviewed papers, Workshops, and short courses will serve to document your work and to make it available to others. Service contributions can come from training others in your department or college or in other institutions to use the Workshop model. In addition, there are challenging opportunities to train teachers from the local high schools and to have your peer leaders model the process for high school students. You can provide opportunities for your Workshop leaders to make presentations at local and national meetings. All these activities will increase your visibility and impact and those of your department and institution.

To engage in educational reform, you must be able to trust the system that will evaluate and reward you. In order to ensure that your efforts will be adequately rewarded when promotion and tenure decisions are made, you need to prepare a well-developed plan that describes your agenda for teaching, scholarship, and service. Get your plan approved in writing in advance by appropriate administrators. Make sure the guidelines for evaluation of Workshop activities are clearly understood and agreed to by those who will be involved in evaluating your performance.

**Seek Allies**

The barriers to change vary with the goals and with the local context. In some departments and institutions, faculty are free to make structural changes in their courses. In others, course changes must be negotiated with committees and administrators. All teachers must be prepared to explain and justify change to their students. If the goals transcend particular courses and aim for departmental or institutional change, then the barriers will be even higher. Regardless of the details, all innovation will meet resistance, and all innovators can use help to get over the barriers.

All the traditional methods of building support will help. Identify like-minded colleagues. Explain the Workshop Project to them. Try to convince some of them to try it. Try to get the department chair and the dean to back up your Workshops. Do not tell your colleagues that Workshops are better. Tell your colleagues how Workshops helped solve problems that faced you in your work and how they might help with the problems confronting your colleagues in their work. If you cannot identify the problem that will motivate change, do not bother to try to win support. You will not even get anyone's attention.

There may be a number of nontraditional sources of support. The innovator needs to seek out these potential allies and bring them into the Workshop Project. The way to do this is to identify individuals and organizations with problems and goals that are the same as yours or complementary to yours. Look for synergy. A classic example is a learning center. Most colleges have some kind of structure that is designed to help students learn; its charge is similar to yours. This office will usually have skilled personnel, organizational resources, and a budget. All those could help you in the design and implementation of Workshops for your course. This same office may be frustrated because its efforts to help students are not connected to faculty or courses. This is a perfect opportunity for some mutual back-scratching. By sharing your Workshop Project with the learning center, you give them connection and access. In return, they can give you resources and help make the project work. In the best outcome, the faculty and the staff of the Learning Center become a team to train Workshop leaders and to manage the operation. In some cases, Workshops may even offer the learning center a better way to spend their budgeted funds.

Support for Workshops can come from surprising sources. In one case, an institution was awarded scholarship funds to "help students learn." An imaginative administrator recognized that the Workshop Project had the same goal and put the two together, with one program funding the other. Industry has identified specific needs for communicative team players, and the Workshop Project aims to teach those same skills. A good faculty member and a good development officer ought to be able to put the two together. Any opportunity to incorporate the Workshop Project into a grant application is a sure-fire way of gaining legitimacy in the eyes of colleagues and administrators. Even better, of course, the grant might get funded and provide real financial support for innovation.

A different source of allied support can be found in the external community involved in the Workshop Project. By definition, this consortium of faculty, staff and students is committed to the development, implementation, and propagation of the Workshop model. Within this group, there is a variety of expertise in a variety of contexts. The support can range from therapy to theory to tactics to technique, all freely available. Just ask.

Finally, we have saved the best for last. We speak of the Workshop Project as a coalition of faculty, learning specialists and students. As we have argued above, the coalition should include administrators as well, but the most important allies for change are the students themselves. Their power is extraordinary. It derives, in part, from their numbers; in part from their influence (e.g., financial or legislative); in large part from their drive, energy, and capacity for change. Above all, their power is primordial, an ur-power. They are the reason that we have colleges and universities.

The Workshop leaders are superb allies, precisely because of their leadership qualities. They are wonderful students and delightful individuals, full of energy, talent, and accomplishment. They are outstanding examples of what most students want to be: individuals who can do good work and have fun doing it. Within the Workshop context, they are irresistible role models for the students, credible cheerleaders for the subject and the course, and nonauthoritarian guides to learning organic chemistry. Beyond the Workshop context, they emerge as leaders in the student body, in their major departments, and in the broad set of curricular and extracurricular opportunities available to them. Their interest in teaching is often kindled by the Workshop Project, and they appear regularly as assistants in other courses and laboratories. Most importantly, they become persuasive agents for change in other courses and departments. In this capacity, they can make observations and suggestions that would be perceived as meddlesome if they were to come from faculty members.

Students have very good ideas about what helps them learn and what does not. After all, they are in the thick of trying to learn. When the students report that an innovation works to help them, it behooves the faculty and the institution to pay attention. When they say that they also like the new structure, then the demand to pay attention is even stronger. Because the leaders are especially sophisticated learners, their views about Workshops carry special weight. Each Workshop reaches six to eight students; that is, the Workshop is an amplifier for the Workshop model of learning. The end result is a large number of students who understand the model and have benefited from their participation.

There is yet a different way in which the Workshop Project builds alliances between faculty and students. The Workshop structure is designed by faculty to help students learn. So are other structures, but the success of the Workshop is absolutely dependent on the work of the students. No set of Workshop problems will do the job if the students do not join as active, prepared individuals, committed to one another's' growth and learning. A lecture might be judged successful, in the absence of students. The parallel proposition for Workshop makes no sense whatever. By the simple act of sharing the work of learning with the students, the Workshop structure radically changes the faculty-student dynamic. By making the student a participant, the Workshop recognizes and supports the student as a unique individual. The effect of making this a faculty-student-institutional-alliance for learning is profound and revolutionary. It is the fundamental logic of the Workshop Project.

**Your Job Is to Teach Well; the Institution's Job Is to Help You**

It is easy to get roles mixed up in academic settings; the result is confusing and usually impedes progress to the desired outcome. The primary task of the faculty is to define, develop and implement methods to help students learn. The primary job of the managers is to help faculty reach that goal. In the context of the Workshop Project, the role of the faculty is to develop the model, to explain it to others and, in self-interest, to identify barriers to implementation. All too often we get tangled up in trying to find ways to knock down the barriers. Administrators, on the other hand, have the power to do just that. In fact, they have been chosen because they have skills, interest, and experience in overcoming obstacles and making things happen. We should play our role and let them play theirs. At most, faculty can contribute to creative discussions about different ways to solve the problems that have been identified. Too often, however, we make the error of presenting solutions and insisting that administrators carry out our solutions. A much better plan is to show how Workshops can help with problems that are shared by faculty and the institution. The faculty objectives should be to secure the cooperative energy and resources of the managers in service of a common cause.

**Be Honest about Costs (and Benefits)**

***Costs.*** Universities and colleges are struggling with the challenge to "do more with less." In the least sophisticated corners of the academic world, this translates into a plan to increase the size of the batch that is processed; that is, increase class size and teach more students with fewer faculty. In other corners, there is enthusiasm and optimism about technological innovations; that is, use machines to teach more efficiently. Both of these approaches derive from a manufacturing model of higher education. The model is deeply flawed, of course, because the products of higher education are not widgets.

A more sophisticated method for doing more with less is to improve quality. This model is perfectly appropriate for higher education because it is coincident with the fundamental ambition of higher education to improve the quality of thought, purpose, and action. The challenge is a bit more acceptable if it is rephrased to "do better with less." Even so, the problem lies in the implementation. Faculty are exhorted to do a better job, but are frustrated by the lack of resources. Administrators want to lead but are hard pressed to find resources for good initiatives. It is easy for administrators to think of faculty as the eternal poor, forever appearing with tin cups in hand. In turn, it is easy for faculty to accuse administrators of trying to get something for nothing. The productivity challenge has the makings of a standoff. In this section we will first analyze costs in the traditional manner: we will itemize the resource costs associated with a unit of Workshop instruction. In the most important part of this section, we will analyze resource costs in terms of a unit of student learning. Although Workshops may seem to be expensive in the "instruction model," they turn out to be a bargain in the "student learning model."

In the traditional model, the Workshop leaders are considered to be part of an instructional delivery system. In most situations, the leaders receive a stipend of the order of $500/Workshop/semester; let's make it $520 for convenience. If the average size of a Workshop is eight students, the cost is $65/student/semester. If there are 13 Workshops/semester, the cost is $5/student/Workshop. If each Workshop is two hours long, the cost is $2.50/student/hour. In the context of today's tuition charges, this does not seem like an exorbitant expense. On the other hand, it is a new expense. In days of tight budgets, the money must come from not paying for some preexisting program, and it is an expense that adds up. For a class of 100 students, the expense for leader stipends at the stated rate is $6500/semester. Of course, tuition income also increases according to the number of students.

The other significant costs associated with the Workshop Project correspond to the allocation of faculty and staff time to the project. Just how these costs are treated depends on the local situation. If there is unused capacity or capacity that is being wasted in unproductive activities, then there may be no new faculty/staff costs associated with a new Workshop Project. The more likely scenario, of course, is that faculty and staff are fully engaged in their current activities. The new Workshop Project imposes new expenses either for replacement or release time or reallocation costs associated with giving up one activity in order to do another. The following are essential time and energy costs associated with the Workshop Project:

1. leader training
2. preparation of Workshop materials
3. organizational issues
   a. hiring and paying leaders
   b. scheduling spaces and times for Workshops
   c. organizing students and leaders into Workshop units
   d. monitoring Workshops in progress
4. evaluations and assessments

Modest costs for materials are associated with these activities. In some situations, some computing equipment may be required.

Much of this book is designed to help new Workshop Projects minimize these time and energy costs. Models and materials are provided to decrease the startup costs for new Workshop Projects, and some of the costs are first-year costs only. Others are steady-state costs. Preparing Workshop materials is mainly a first-year project; the leader training goes on every year. The partitioning of the work will also depend on the local situation. Workshop materials obviously belong to the faculty, and as we have discussed elsewhere, the faculty must be involved in the leader training. However, the distribution of leader training work between faculty and learning specialist will vary. The same is true for organizational and assessment issues. In general, the faculty may do it all for small classes; faculty teaching large classes will need supporting help, but that is to be expected.

The following is a *maximum* budget for a one-semester class of 160 students, assuming that the faculty have strong support and all costs are incremental:

| | | |
|---|---|---|
| 1. | 20 leader stipends @ $500 (8 students/leader) | $10,000 |
| 2. | Release time for learning specialist to teach leader training class | 2,500 |
| 3. | Supplies, copying | 1,000 |
| 4 | Staff support | 1,500 |
| 5. | Computing hardware/software | 500 |
| 6. | Evaluation | 500 |
| | Total | $16,000* |

*Resource cost/student = $16,000/160 = $100

There are two obvious ways to control the new costs for Workshop leader stipends. The first is to reduce the leader stipend, in the limit to zero. Our view is that the compensation sets the stage for a contractual, professional relationship between the Workshop Project and the leaders. The leaders are in for hard work, personal challenge, and responsibility. They will be generous with their time and effort, just as faculty and learning specialists are generous, but there is no reason for the leaders to work for free. We do not. Having said that, there is room for variation in the amount of compensation. If the leaders are paid, the appropriate amount is best determined by the local competition. What do students get paid for other jobs? What other jobs are available? The other obvious approach to manipulating the leader costs is to increase the number of students per Workshop, thereby reducing the cost/student. We have already discussed that approach in the introduction to this chapter. If the costs/student are halved by doubling the number of students in a group, the result is not a Workshop, and the benefits are lost.

Colleges and universities have been inventive in identifying sources of income to cover Workshop leader costs. Some possible funding sources are:

- college budget
- department budget
- learning center budget
- Work-Study programs
- other student support, office budgets; for example, women in science, underrepresented minority student support programs, supplemental instruction budgets, tutorial programs
- scholarship funds; for example, merit scholarships to leaders
- scholarship funds to support the development of certain groups of students; for example, underrepresented minority students, women
- institutional initiatives and programs to support teaching reforms
- government, corporate, and foundation grants to support teaching reform
- alumni annual giving for good teaching, and development office activities with individual donors

Some universities may be tempted to roll over graduate-student TA budgets to support undergraduate Workshop leaders. Since graduate students are more expensive, the result looks like cost savings. Even if that thought has never occurred to deans, the faculty are certain that it will. As a result, faculty in graduate institutions fear that the idea of peer leadership will undermine the *raison d'être* for the TA budget and, therefore, the budget and the graduate program itself. This fear can sink a Workshop Project before it is launched. Some of this anxiety is misplaced and derives from the incorrect assumption that a Workshop leader is a mini-TA, at lower pay. As we have discussed, this is absolutely incorrect. The Workshop leaders have different skills and roles from those of the graduate students. In fact, the Workshop model provides an appropriate role for undergraduate peer leaders that is perfectly consistent with their education and experience. When undergraduates have been used in place of graduate TAs, they have often been cast in roles that they are not prepared to play.

Our experience is that graduate students make excellent Workshop leaders. They must be trained for the new role, of course. In general, graduate students like the Workshop model because the interactions with the students are active and satisfying. Because the Workshop and the training are carefully structured, they know what they are supposed to do when they get to Workshop. Our experience is also that large courses continue to need graduate-student expertise. In general, we keep the undergraduate peer leaders out of grading/evaluation situations. They do not yet know enough chemistry to be the best graders, and their role is to support and encourage, not to evaluate. The graduate students are essential to help with the grading, to hold office hours, to supply expert chemistry help, to conduct weekly review sessions, and so on. At the University of Rochester each graduate student associated with the organic course does two two-hour Workshops per week and grades every exam. This workload is considered by the graduate students to be comparable to a one afternoon per week laboratory TA assignment. The graduate-student support budget for the organic course at Rochester is exactly the same, pre- and post-Workshop. So far, no one has tried to reduce the graduate TA budget because of the Workshop Project.

***Benefits.*** It is essential to be honest about the costs in dollars and in time and energy for a Workshop course; it is also important to be clear about the benefits. The traditional analysis of costs is lopsided because it deals only with the allocation of institutional resources to the instructional work. Although this work may get normalized to a per student basis, the benefit to the student does not appear in the reckoning. The resource cost per student is the same whether all students fail or all earn A's.

What if we were to incorporate the student outcomes into the model? Then it would matter a lot whether students got A's or E's. In the most naive kind of calculation, we might divide the resource cost per student for a given course by the grade point average for the students in that course. The resulting resource cost unit of student learning would be smaller if all students learned at the A level (GPA = 4). In the limit that they all failed to learn (GPA = 0), the cost/unit of learning would approach infinity, as it should. A course that leads to that result is obviously inefficient and unproductive.

The naive approach of using GPA to measure student learning is obviously flawed. There are too many assumptions hidden in the phrase "A-level learning," but the example serves to make the point that resource cost per student is not the same as the resource cost per unit of learning. Colleges and universities use the resource cost per student model because they have developed metrics for counting the instructional costs. We obviously need to develop metrics for counting units of learning.

The GPA for the class, however, is not a completely silly way to measure student learning. It is the "coin of the realm," and there is an implicit assumption that it correlates with learning. Clearly, there are important questions about equivalent standards when GPAs are compared, but we compare GPAs all the time. "Student success," defined as the percentage of students who complete the course with C or better, compared with all students who were in the course, is a different measure of student learning. Other measures of units of learning could be developed. It might be hard to work out these new metrics, but that is beside the point. A measure of productivity that does not incorporate a measure of outcome is far worse than one that incorporates a somewhat flawed measure. The former is dead wrong; the latter just needs an adjustment.

The data on the effects of Workshops on learning are compelling and statistically significant. Workshops improve student learning, as measured by GPA and by "student success." In short, students like Workshops and do better. In addition, attendance is high and students report that Workshops are fun. In a broad survey ($n = 723$) across twelve different courses, teachers, and institutions, 82% of the students reported that they would recommend Workshop courses to others. Since the student's affective response to college produces multiple effects on learning, the changes in student satisfaction may have significant consequences for the resource cost per unit of learning.

The Workshop leaders are a very reliable source of information, because they are among our best students. Across the same twelve courses, they ($n = 100$) agreed that Workshops improve students' grades (91%) and that they would recommend Workshops to others (100%). From a different perspective, the opportunity to be a Workshop leader provides recognition and reward for our outstanding students. Even though they had already done A-level work in their chemistry course, the leaders reported that "acting as a leader increased my understanding of chemistry"; 98% agreed. The competition for leader positions is keen. In part, students recognize that they are being chosen to lead their peers because of their own accomplishments and talents. We suspect that there are many other positive consequences for the leaders; for example, new relationships with faculty and the institution; new understanding of their own learning processes; new career aspirations; new communication, team, and leadership skills; increased confidence; and a beginning sense of professional responsibility. We are working hard to test these hypotheses. Regardless of the details, all the leaders report that they learned a lot.

It is very hard to measure just how much the increases in student learning are worth. It is even harder to put a number on the value of improved student success or on the many dimensions of leader learning, but that does not mean that we should ignore the effects. It simply means that our estimates of value should have large standard derivations of the mean. Our best guess is that the cumulative effects on learning are very significant; that is, the Workshops produce many units of learning. *As a result, the resource cost per unit of learning is far less than the resource cost per student.* The actual resource cost/unit of learning from a Workshop may be a bargain in our present scheme of higher education. We do not have good comparative figures for the resource costs per unit of learning for other educational investments. Our guess is that Workshops are much less expensive.

Finally, it is important to recognize that training is reversible (i.e., forgettable), but learning is not; that is, learning new patterns of thought and standards of understanding is like learning to ride a bicycle. If Workshop facilitates learning, as opposed to training, then the impact of the investment continues forever. The units of learning compound over time, and the resource cost per unit of learning ultimately approaches zero!

Lest this seem like so much numerical double-talk, it is important to consider the sources of these "magical" gains in productivity. One useful line of thought is to recognize that the Workshop engages the power of the students to learn by interacting with one another. If learning corresponds to a process of constructing internal understanding, then the interactions in the Workshop facilitate this process. The productivity gain comes from transforming a natural resource into units of learning. Because the power is free, the effect only appears magical. In contrast, the compounding is magic, as noted by Einstein.

In order to close the circle on this discussion of costs and benefits, we need to return to the initial observations. Although the resource cost per unit of learning may be an extraordinary bargain, it is a *new cost*, and it adds up. In the short run, there is no escaping the need to find the funds to support the Workshops. In the long run, there is every reason to believe that improvements in quality will be rewarded with improvements in revenues. Just how this happens will depend on the institution. An improvement in retention brings an immediate return in tuition revenue. In the longer run, improvements in learning and success means that the institution is more productive and therefore worthy of increased public support or a higher tuition. If students believe that their struggles to learn have been acknowledged and supported by the institution, they will be more likely to return support in the subsequent years. In the short run, the institution needs to have the courage to invest in a program that will bring a high rate of return in the long run.

## References

Barr, R. B. and J. Tagg (1995). From Teaching to Learning - A New Paradigm for Undergraduate Education. *Change 27*: 13-25.

The Boyer Commission on Educating Undergraduates in the Research University (1998). Reinventing Undergraduate Education: A Blueprint for America's Research Universities. http://notes.cc.sunysb.edu/Pres/boyer.nsf/

Boyer, E. L. (1990). *Scholarship Reconsidered: Priorities of the Professoriate.* Carnegie Foundation for the Advancement of Teaching, Princeton, N.J.

Coppola, B. P. (1996). Progress in Practice: The Scholarship of Teaching. *The Chemical Educator 1*, no .3, http://journals.springer-ny.com/chedr

Fiesel, L. D. (1995). Part of Faculty Rewards: Can We Implement the Scholarship of Teaching? An On-Line Symposium. http://www.inform.umd.edu/EdRes/Topic/Chemistry/ChemConference/Faculty Rewards/home.html

Glassick, C. E., M. T. Huber, and S. I. Maeroff (1997). *Scholarship Assessed.* San Francisco: Jossey-Bass.

Guskin, A. E. (1994). Restructuring the Role of Faculty. *Change 26*: 16-25.

Moore, J. W. (1997). Editorial: Scholarship in Chemical Education. *Journal of Chemical Education 74*: 741.

Paulsen, M. B. and K. A. Feldman (1995). *Taking Teaching Seriously: Meeting the Challenge of Instructional Improvement.* Eric Digests, ERIC Clearinghouse on Higher Education, Washington, D.C. ERIC Document 396615.

Seymour, E. (1998). *Tracking the Process of Change in U.S. Undergraduate Education in Science, Mathematics, Engineering, and Technology.* Paper presented to the 1998 International Gordon Research Conference on Science Education: "New Developments and Visualization in Chemistry and Science Education," held at Queen's College, Oxford, U.K.

Schon, D. A. (1995). The New Scholarship Requires a New Epistemology. *Change 27*: 27-34.

Shulman, L. S. (1993). Teaching as Community Property-Putting an End to Pedagogical Solitude. *Change 25*: 6-7.

# *Chapter Six*
# *Workshop Evaluation*

**Leo Gafney**

## Background

The evaluation of Peer-Led Team Learning has been an integral part of the Workshop Project. The evaluator regularly brought data, findings, and conclusions to the faculty members implementing the model. A number of these faculty participated in the evaluation by using questionnaires, arranging interviews, and discussing the program, but more importantly, the faculty themselves planned and carried out comparison studies that convinced them and their colleagues that PLTL is effective.

The evaluation process led to the development of the *Critical Components*, and these in turn have provided a model for the program as it is adopted, adapted, and implemented at different sites and for different disciplines. When there were serious difficulties in implementing PLTL, problems always surfaced in one or another of the *Critical Components*. Thus, the *Critical Components* can be used as an effective structure for evaluation.

This chapter is divided into the following sections.



## Evaluation Overview

***Startup.*** Focus groups conducted early in the project revealed that students and student leaders were enthusiastic and reflective about the PLTL model and the way in which the Workshops complemented other course components. Students commented that professors sometimes assumed knowledge and "started in the middle" or "took shortcuts." They wished that the professors would "stay on a topic a bit longer." These difficulties with lectures were nicely compensated by the Workshop. The leaders, they said, were able to explain things "in a different way." They "knew where you were coming from."

Students and student leaders in focus groups also said that the PLTL environment permitted the students to try different approaches, to make mistakes, to test their understanding. These activities are clear signs of active learning. Students put themselves into the process, were willing to take chances, and saw themselves benefiting from the interactions.

***Cooperative learning.*** There is a large body of literature confirming the effectiveness of collaborative learning. Studies of Treisman-like Workshops (Gillman 1990; Treismann 1992) as well as extensive reviews of the research (Johnson and Johnson 1989; Johnson, Johnson, and Smith 1991; Slavin 1983) indicate that small group work enhances problem-solving skills, builds interest and confidence, and prepares students for a world of work in which communication and team skills are essential for success. Johnson, Johnson, and Smith (1991) point out that "We know far more about the efficacy of cooperative learning than we know about lecturing, departmentalization, the use of technology, or almost any other facet of education." Halpern (1994) provides useful information, data, and guidelines about the use of cooperative learning at the college level.

The principal investigators, evaluator, and professors adopting PLTL were aware of the successes of collaborative learning and extracted lessons appropriate to the model. Decisions were made about group size, learning activities, and the development of appropriate materials. These provided a context for continuing evaluation.

Most forms and models of collaborative learning do not include the role of a permanent leader. Roles are often assigned or emerge in group activities, but for PLTL, *the peer leader is an indispensable element in the group process*. Consequently, the role and activities of the leader became important in the evaluation.

***Evaluation plan.*** The evaluation plan was designed to be flexible, allowing for revision as new data and insights emerged. The plan evolved in the course of the project, including both formative and summative strategies. In its first phase the evaluation considered how PLTL was being implemented in different locations. It had been assumed that there was a standard format that included the following: lecture time diminished by at least one hour; Workshops of two hours; groups containing six to eight students; leaders trained in some form of cooperative learning; professors directly involved with the supervision of the Workshop leaders. It turned out that one or several of these conditions were often absent. There were, however, enough common elements to permit an evaluation across the various sites. These common elements led to the articulation of the *Critical Components* with added detail.

The second phase of the formative evaluation used a revised survey, extensive phone interviews, and observations, all based explicitly on the *Critical Components*. As expected, the *Critical Components* served as excellent indicators, often proving insights about why certain program activities were not working as well as expected.

In addition, professors took the initiative to measure student performance and compare the outcomes of Workshop and non-Workshop courses. The data generated by these studies are summative; they provide clear evidence that the Workshop students perform better than non-Workshop students when comparisons are made using traditional classroom measures. This work is continuing.

***Formative findings.*** The several phases of the formative evaluation provided solid data for the following findings.

- The *Critical Components* provide a workable model for implementing and evaluating PLTL. Care is needed regarding the details of each component.

- Within the framework of the model there is ample room for adaptation according to local needs.
- Students overwhelmingly find PLTL helpful to their learning when implemented according to the *Critical Components*.
- The peer leader distinguishes PLTL from other forms of cooperative or team learning. The training and supervision of the leaders is central to the program.

Greater detail on formative findings and conclusions can be found in the section on *Critical Components*, after Formative Evaluation 1, in this report.

***Summative findings: student performance.*** Grades comparing Workshop and non-Workshop students have been gathered from a number of professors. Methods of establishing and comparing groups varied widely according to the local situation. Three professors had classes divided into comparable groups. Others compared different classes taking the same course. Some compared grades of current Workshop with past non-Workshop classes. The data show a difference in the percentages of A, B, and C grades in favor of the Workshop students for almost all groups compared. A number of comparative studies are being conducted as PLTL is extended to additional sites in different disciplines. These will add to the body of data on the method.

We hope that this section of the guidebook will assist in the preparation and implementation of evaluation studies. A number of questions remain about the teaching methods and approaches to student assessment most appropriate for use in connection with the Workshop goals and activities. These provide opportunities for new research about PLTL.

## Formative Evaluation 1: 1996 Survey

***Surveys.*** During Fall term 1996-1997 written questionnaires were sent to the participating institutions to obtain data about students' experiences in the Workshop courses. Students were asked about relationships with the Workshop leaders and with other students, involvement of the faculty, and the materials and arrangements used for the Workshops. Workshop leaders were asked similar questions, also about their training and support in conducting the Workshops, and whether they viewed the Workshops as generally successful.

Nine institutions and at least sixteen different classes were represented in the survey. Responses were tabulated according to institutions. The rate of student sample was generally greater than 50% and frequently nearly 100% of those in a Workshop course.

***Returns.*** Table 1 shows the numbers of responses from students and student leaders according to twelve groups of students and nine groups of peer leaders. Responses to the survey questions were tabulated as *agree*, *neutral* or *disagree* for each group. Each group was thus considered as a separate sample. The 1996 survey questions and the detailed responses are reported in the charts at the end of this section. The highlights of the results are presented here. Complete, ready-to-use, revised forms of these student and leader surveys are included in Appendix III.

***Students.*** The student surveys were analyzed by group averages. In addition, the averages of the groups were averaged to give overall averages. In general, the differences between groups and overall averages were slight. The overall averages are reported in the chart.

In general, students reported that Workshops were helpful to their learning. Items 3, 4, 9, 10, 15, and 16 discuss learning and were all agreed to by more than 75% of the respondents, whereas less than 10% disagreed with any of these items. These items contain statements that

**Table 1. Survey Groups and Responses**

| Institution | Course | Number of Responses Student | Leaders |
|---|---|---|---|
| City College | Gen Chem 104 | 102 | |
| | Gen Chem 103.1A | 105 | |
| | Gen Chem 103.1B | 21 | |
| | Gen Chem 103.1C | 34 | 7 |
| University of the Pacific | Organic | 28 | 4 |
| University of Pittsburgh | Gen Chem | 115 | 14 |
| St. Xavier University | Organic | 25 | 5 |
| NYC Tech | Gen Chem | 34 | 6 |
| University of Rochester | Organic | 166 | 25 |
| University of Pennsylvania | Organic | 30 | 3 |
| Medgar Evers | Gen Chem | 46 | 6 |
| Borough of Manhattan Community College | Gen Chem | 17 | 5 |
| **Total** | | **723** | **75** |

interacting with the Workshop leader helps learning; the materials are demanding, integrated with the course, and useful; that students would recommend the Workshop course; students are comfortable asking questions; and the leader is well prepared. Responding to "Workshops are improving my grade," nearly 70% of the students agreed, and only 13% disagreed.

There was rather wide variance among group averages to some items. Average agreement with the statement that "the course as a whole is well organized" ranged from 50% to 91%. Average agreement with the statement that "the lecturer clearly presents the chemistry" ranged from 37% to 100%. Many students who found that the course was not well organized or that the professors were not clear nevertheless reported very favorably about the positive impact of PLTL on their learning.

***Leaders.*** The leader surveys were analyzed by group averages. In addition, the averages of the groups were averaged to give overall averages. In general, the differences between groups and overall averages were slight. The overall averages are reported in the chart.

As in the case of the students, the leaders' responses indicate highly positive experiences about the course and the Workshops. Agreement with "I would recommend Workshop courses to other students" is even greater (100%) than in the case of students (82%). The leaders also agreed, by a wider margin than the students, that the Workshops improve student grades: leaders, (91); students (70%). The leaders are much more involved and have a greater emotional investment than the students. Consequently, it is natural that they may consider the Workshops more important than students do.

Regarding "Acting as a Workshop leader increases my understanding of chemistry," 98% agreed, and none disagreed. Responding to "I act more as a guide than a teacher," 76% of the leaders agreed. Similarly, "leader training was useful" received 81% leader agreement.

Both surveys state that "uninterested and unmotivated students make it difficult for others to benefit from Workshops." The students disagreed with this more often (59%) than leaders disagreed (40%).

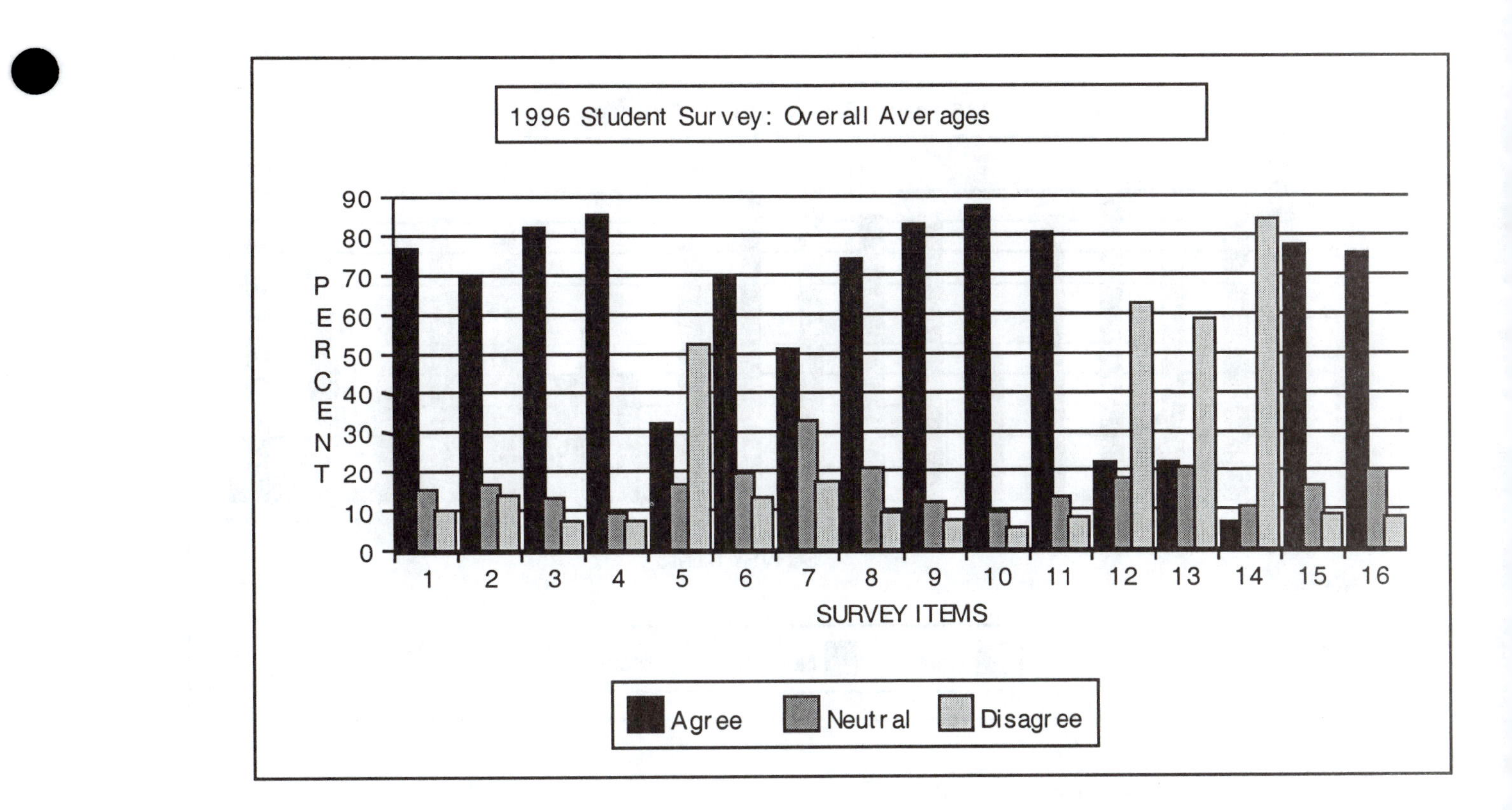


**Student Survey: Fall 1996**

| | | Disagree 1 | 2 | 3 | 4 | Agree 5 |
|---|---|---|---|---|---|---|
| 1. | The course as a whole is well organized. | 1 | 2 | 3 | 4 | 5 |
| 2. | The lecturer clearly presents the chemistry. | 1 | 2 | 3 | 4 | 5 |
| 3. | Interacting with the Workshop leader increases my understanding of chemistry. | 1 | 2 | 3 | 4 | 5 |
| 4. | The Workshop materials are well connected to the lectures. | 1 | 2 | 3 | 4 | 5 |
| 5. | My Workshop group sometimes has extra meetings to prepare for tests or to review difficult material. | 1 | 2 | 3 | 4 | 5 |
| 6. | I believe that the Workshops are improving my grade. | 1 | 2 | 3 | 4 | 5 |
| 7. | I regularly explain problems to other students in the Workshops. | 1 | 2 | 3 | 4 | 5 |
| 8. | Interacting with the other group members increases my understanding of chemistry. | 1 | 2 | 3 | 4 | 5 |
| 9. | I would recommend Workshop courses to other students. | 1 | 2 | 3 | 4 | 5 |
| 10. | In the Workshops I am comfortable asking questions about material I do not understand. | 1 | 2 | 3 | 4 | 5 |
| 11. | The lecturer encourages us to participate in the Workshops. | 1 | 2 | 3 | 4 | 5 |
| 12. | Noise or other distractions made it difficult to benefit from the Workshops. | 1 | 2 | 3 | 4 | 5 |
| 13. | Students who are uninterested or unmotivated made it difficult for others to benefit from the Workshops. | 1 | 2 | 3 | 4 | 5 |
| 14. | The Workshops are not helpful because I already know almost everything that is covered. | 1 | 2 | 3 | 4 | 5 |
| 15. | The Workshop materials are demanding and are good preparation for the tests. | 1 | 2 | 3 | 4 | 5 |
| 16. | The Workshop leader is well prepared. | 1 | 2 | 3 | 4 | 5 |

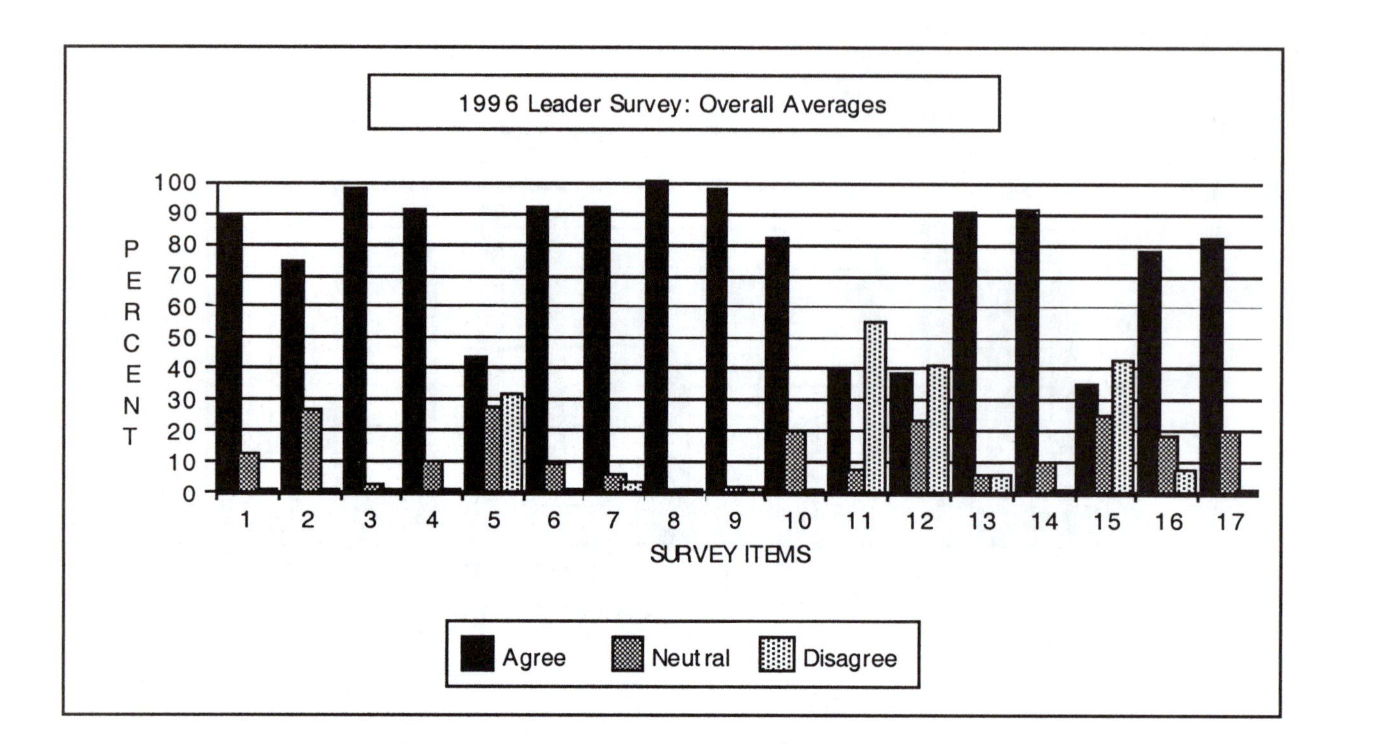


**Leader Survey: Fall 1996**

| | | Disagree 1 | 2 | 3 | 4 | Agree 5 |
|---|---|---|---|---|---|---|
| 1. | The course as a whole is well organized. | 1 | 2 | 3 | 4 | 5 |
| 2. | The lecturer clearly presents the chemistry. | 1 | 2 | 3 | 4 | 5 |
| 3. | Acting as a Workshop leader increases my understanding of chemistry. | 1 | 2 | 3 | 4 | 5 |
| 4. | The Workshop materials are well connected to the lectures. | 1 | 2 | 3 | 4 | 5 |
| 5. | My Workshop group sometimes has extra meetings to prepare for tests or to review difficult material. | 1 | 2 | 3 | 4 | 5 |
| 6. | I believe that the Workshops improve student grades. | 1 | 2 | 3 | 4 | 5 |
| 7. | I regularly explain problems to students in the Workshops. | 1 | 2 | 3 | 4 | 5 |
| 8. | I would recommend Workshop courses to other students. | 1 | 2 | 3 | 4 | 5 |
| 9. | In the Workshops students are generally comfortable asking questions about material they do not understand. | 1 | 2 | 3 | 4 | 5 |
| 10. | The lecturer shows an interest in me as a Workshop leader. | 1 | 2 | 3 | 4 | 5 |
| 11. | Noise or other distractions sometimes make it difficult to benefit from the Workshops. | 1 | 2 | 3 | 4 | 5 |
| 12. | Students who are uninterested or unmotivated make it difficult for others to benefit from the Workshops. | 1 | 2 | 3 | 4 | 5 |
| 13. | Interacting with the other Workshop leaders is helpful. | 1 | 2 | 3 | 4 | 5 |
| 14. | The Workshop materials are demanding and are good preparation for the tests. | 1 | 2 | 3 | 4 | 5 |
| 15. | Students are generally well prepared for the Workshops. | 1 | 2 | 3 | 4 | 5 |
| 16. | As a Workshop leader I act more as a guide than a teacher. | 1 | 2 | 3 | 4 | 5 |
| 17. | The training that I have or am receiving on how to conduct Workshops is helpful | 1 | 2 | 3 | 4 | 5 |

### *Critical Components*: Detailed Analysis

Initial summative results show that peer-led team learning enhanced student learning. The next phase of evaluation, essential for dissemination, was to consider the *Critical Components* required for success. The years of evaluation activities have added considerable detail to the role and importance of the *Critical Components*.

- **The organizational arrangements, including the size of the group, space, time, noise level, teaching resources, and the like promote learning.**

- **The Workshop materials are challenging at an appropriate level and, integrated with the other course components, intended to encourage active learning and to work well in collaborative learning groups.**

- **The peer leaders are students who have successfully completed the course. They are well trained and closely supervised, with attention to knowledge of the Workshop problems, teaching/learning strategies, and leadership skills for small groups.**

- **The faculty teaching the courses are closely involved with the PLTL Workshops and the peer leaders.**

- **The PLTL Workshop sessions are integral to the course, coordinated with other elements.**

- **The institution, at the highest levels of administration and pedagogy, and at the departmental levels, encourages innovative teaching and provides sufficient logistical and financial support.**

***Organizational arrangements.*** There are a number of important Workshop arrangements, particularly space, time, and the size of groups. The Workshop model recommends

- a two-hour Workshop, held once a week, with about six to eight students;
- that attendance be required; and
- that space be adequate for concentrated small-group activities.

Surveys and observations indicate that at shorter Workshops less time is devoted to group activities, conceptual work, and guided problem solving.

***Materials.*** In general, professors adopting the Workshop approach spend considerable time writing or adapting materials. The model recommends

- materials that are engaging and challenging but not so difficult as to discourage students;
- materials that develop skills and knowledge directly related to tests and grades; and
- materials appropriate for small group work.

Materials go a long way toward establishing the success of the Workshop and the kinds of interactions among the group and the peer leader. Often a pre-Workshop activity is used to prepare students. Interviews with professors reveal a general appreciation of materials made available by the Project, combined with a need to adapt materials to local situations. The same materials have been reported to be too difficult at one site and too easy at another.

The materials prepared, piloted, and published by the Workshop Chemistry Project have been useful at a number of sites; the process can provide a model for other disciplines.

***Leaders.*** The training of peer leaders varies from site to site depending on the interest of the professor and the involvement of a learning specialist or someone trained in science and math education. Efforts are still being made to determine the *Critical Components* of leader training. At the present time the model recommends that

- Workshop leaders be skilled in working with individuals and with small groups;
- they perform as facilitators rather than lecturers or TAs;
- they have some training before they begin, and continuing training as they work as peer leaders; and
- they be competent in their knowledge of the discipline and as problem solvers.

The presence and activity of the peer leader distinguishes the Workshop from many other varieties of cooperative learning. Peer-led teams are essential to the PLTL model. Selection of leaders is usually based on grades in the course and interviews. Training varies from a few hours to several days. Several institutions require a for-credit course for the leaders. Credit is given at some sites in education, at some sites in science. Students meet regularly, discuss the Workshop materials, and the dynamics of learning, maintain journals, and share experiences. Regardless of the details, all agree that the leaders must be well prepared in several areas: knowledge of the chemistry and problem-solving skills, ability to organize and direct group learning activities; and familiarity with approaches to student problems ranging from difficulty with the course to personal adjustment

***Faculty.*** Professors adopting the PLTL model are involved at differing levels concerning the supervision of leaders, attendance at the Workshops, and the development of materials. Guidelines for the most appropriate levels of faculty involvement exist in different parts of program documents but have not yet been codified, and there are a number of open questions. The PLTL model recommends that the professor

- prepare, review, and update the Workshop materials;
- preview the Workshop materials and activities with the student leaders; and
- be available for consultation with the peer leaders and with students.

It sometimes happens that the management of the Workshops is taken on by a member of the faculty not teaching the course or by a graduate student. This process generally leads to problems. First, the Workshops are most successful when they treat material from that week's lectures. Much is lost if Workshops are not coordinated with the lectures. Similarly, students benefit much more if the professor's familiarity with the Workshops leads to references to them as expected and important parts of the learning process.

***Integration.*** This is an umbrella category bringing together the other components. Also, we have observed an important morale factor. Students report positively about the Workshops when their experiences in the Workshops contribute to their success in the course.

- The Workshops take a considerable amount of student time and energy. Consequently students must value the Workshops, or the participation and impact will be considerably diminished.
- Second, integration means that the leaders are aware of the approach taken in the lectures, and the professor's overall method.
- Third, the PLTL model requires that the professor refer to the Workshops in lectures and adjust lectures to make good use of the Workshop time.

***Support.*** This is critical for institutionalization. The Workshop program cannot survive without adequate resources, nor will it survive if implemented by one or two faculty. A critical mass is required for the PLTL approach to take root and become a normal part of the business of a department and institution. The model suggests that

- the Workshop approach be extended across several courses and disciplines;
- administrators such as department heads and deans support the PLTL approach; and
- the institution provide local funding.

Interviews with PLTL faculty have uncovered a pattern regarding institutional support. When first planning Workshops, faculty members are generally enthusiastic about the pedagogical advantages, have acquired some resources with which to pay peer leaders, and have begun to develop materials. Consequently, they are not much concerned with institutional support. But after a few years, particularly if colleagues begin to adopt the PLTL model, the need for on-site support becomes evident and even critical for long-term success and continuity.

***Critical Components analysis.*** Evaluation activities and findings have revealed the need for more detailed analysis of the implementation of the *Critical Components*. A *Critical Component* analysis tests the actual Workshop against the model. We have strong descriptive evidence that the Workshop will not proceed as intended and student performance is less likely to improve if certain elements are omitted. We suggest the following as an aid to self-evaluation.

Table 2 summarizes the detailed components of the Workshop model. The table may be used as the basis for a self-evaluation, assessing the degree to which Workshops at a particular institution conform to the recommended model. The table has been useful in identifying problem areas for PLTL programs that were generally doing well, but not succeeding at the desired level.

**Table 2. *Critical Component* Analysis**

| *Critical Components* | Subcomponents | | |
|---|---|---|---|
| 1. Organizational Arrangements | Time | Space | Group Size Attendance |
| 2. Materials | Fit with course; relate to tests | Engaging and appropriately challenging | Suitable for group activity |
| 3. Leaders | Skill with groups; facilitator rather than teacher | Training and supervision | Discipline knowledge and problem-solving skill |
| 4. Professor's Involvement | Preview of Workshop problems with peer leaders | Preparation and review of materials | Available to students and student leaders |
| 5. Integrated with the Course | Students view Workshops as important to learning | Leaders are aware of lecture approach | Lecturer refers to Workshops |
| 6. Evidence of Support and Growth | Number of disciplines and courses | Administrative support | Local financial support |

## Formative Evaluation 2a: Analysis According to *Critical Components*

***Evaluation methods.*** Faculty members at a number of institutions became familiar with the PLTL model and elected to implement it in their chemistry courses. Most were able to support the project through NSF Adapt and Adopt grants. These new PLTL courses provided opportunities to evaluate the implementation of the *Critical Components* of the model. Activities at these sites were evaluated in Fall 1998 using revised questionnaires for students ($n = 344$) and leaders ($n = 41$) and phone interviews with faculty ($n = 11$) at different institutions. The results of these surveys and interviews are reported according to the *Critical Components*.

***Organizational arrangements.*** The recommended structure for Workshop chemistry is one two-hour Workshop per week with six to eight students and one peer Workshop leader. Although this structure might be considered an average, there is considerable variation at the adapt and adopt sites. The professors interviewed were generally satisfied with the time allocated to the Workshops. Most knew that a two-hour block was recommended but had decided on less for different reasons, most commonly because the Workshop fit a slot previously held by recitation. At least two sites reduced lecture time by about 25% in order to allow for, or to increase, the Workshop time.

We observed a strong correlation between the length of the Workshop session and the nature of the Workshop activities. As the length of the Workshop increases, the time spent on question-and-answer work decreases, and the time spent on group activity increases. This correlation needs to be tested over a larger number of sites. However, the preliminary data suggest that short Workshops should severely limit the number of Workshop problems in order to preserve time for interactive discussion and debate.

| Time | 60 min | 80 min | 90 min | 2 hours |
|---|---|---|---|---|
| Number of Sites | 3 | 1 | 3 | 4 |

The size of the Workshop groups also varied from site to site. Variation seems to be due to several factors: availability of qualified Workshop leaders, or funds for leaders; student attrition; interest in trying a somewhat different model. Sizes of groups at the different sites were reported as follows:

| Group size | 3-4 | 5-8 | 7-10 | 20-30 (4 to 6 subgroups) |
|---|---|---|---|---|
| Sites | 1 | 6 | 2 | 2 |

In response to a statement on the student survey, "Students who are uninterested or unmotivated made it difficult for others to benefit from the Workshop," 52% disagreed, in accord with previous results. Two sites did not have individual group leaders; the Workshop leader and professor floated among the groups. At these sites an average of 39% of the students disagreed with the statement about unmotivated students. At the sites with individual group leaders, an average of 56% disagreed. This suggests that the group leaders may play and important role in maintaining positive dynamics in the groups.

Of the eleven sites represented by faculty in the phone interviews, nine reported that attendance at the Workshops is mandatory, two that attendance is not mandatory but very strongly encouraged. All reported that attendance is very good. This was confirmed by the student leader surveys, which reported an average group size of 7.9 and an average attendance of 6.8 students.

In general we can say that the adapt and adopt sites have made efforts to ensure that the organizational arrangements are adequate for student learning, and that sites are indeed adapting the arrangements with considerable variation according to local need and constraints.

***Materials.*** Professors view the materials as the most important component of the Workshops. The reason for this is straightforward: the materials determine what the students and the leader actually do in the Workshop. Of the eleven faculty members interviewed, six reported that they adapted materials from the those developed by the Workshop Chemistry Project; five said write their own materials. All reported that working on materials was a high priority and a time-consuming activity. Most reported that the materials were at about the same level of difficulty as the problems found in the text. Two said that their Workshop materials were written especially for group work; two said that their materials required more thinking than the text materials; one had an interest in developing materials based on common misconceptions. Most saw the development of materials as an ongoing process that would take several iterations.

In response to the statement, "The Workshop materials are more challenging than most textbook problems," 53% of the students agreed, and 75% of the leaders agreed. In response to an item asking about the sequence of activities in a typical Workshop, almost all the Workshop leaders reported discussing the materials in one way or another: completing the worksheets, doing the Workshop, doing the problems. In response to an item asking about how the materials met different objectives, students and Workshop leaders gave high positive ratings to the materials and were consistent with each other in their ratings. Table 3 shows the responses.

**Table 3. Workshop Materials**

This item is about the materials used in chemistry Workshops. Circle a number from 1 to 5 according to how well the materials meet each objective: **1** = materials do not meet this objective at all; **2** = somewhat meet the objective; **3** = materials meet the objective rather well; **4** = materials meet this objective very well; **5** = materials are excellent for meeting this objective:

| | Percentages of responses | | | | | |
|---|---|---|---|---|---|---|
| | (1-2) | | (3) | | (4-5) | |
| The materials are: | **Ldr** | **Stu** | **Ldr** | **Stu** | **Ldr** | **Stu** |
| (a) integrated with the lecture | 5 | 13 | 33 | 18 | 63 | 68 |
| (b) challenging | 0 | 5 | 23 | 18 | 78 | 76 |
| (c) developed to review fundamentals | 5 | 8 | 15 | 21 | 80 | 70 |
| (d) useful for group work | 0 | 4 | 23 | 20 | 78 | 74 |
| (e) motivational | 30 | 17 | 33 | 33 | 38 | 49 |

Workshop leader responses to the statement, "Are Workshop problems good preparation for tests?" tended to be consistent for a particular institution. About 50% of the respondents gave an unqualified yes." Many included reasons: "problems are from past tests ... development of problem-solving skills ... cementing concepts ... there is direct correlation." About 25% gave a qualified yes to the question, including comments such as: "I think so ... not sure ... in a way ... for the most part." About 25% responded no, with comments such as "students don't see the connection ... not as complicated as test problems."

In summary, the overall pictures is quite clear the Workshop materials are viewed as important and effective by professors, student leaders, and students.

***Leaders.*** In response to a question about selection and training of Workshop leaders, most professors reported that candidates were students who had been successful in the course. The pool from which to select the leaders varied from site to site. Professors who did not obtain an adequate number from among previous students looked to and obtained candidates from other professors, the education department, the supplemental instruction program, mentoring programs. Most said that they considered grades important, but people skills of equal value.

With regard to pedagogical training, seven sites reported an initial orientation meeting of two to three hours; two reported several initial training sessions. One site has a separate two-credit course run jointly by the professor and a colleague from chemistry education. One or two others hope to implement a similar course. When asked, "Were you taught theories of learning and related methods of teaching?" about half of the Workshop leaders responded no; and about half said "a little" or said that they talked about teaching and learning at the weekly meetings with the professors.

Regarding preparation to deal with the contents of Workshops, eight professors reported weekly meetings with the Workshop leaders; the others reported biweekly or occasional meetings. Leaders at one site attend lectures and correct quizzes. Leaders at at least one site maintain journals. When asked about ongoing training in running Workshops, about two-thirds of the student leaders mentioned weekly or regular meetings at which they review the materials and prepare for the Workshops and one-third reported an initial session at the beginning of the course or little or not enough training and support.

In response to a statement indicating comfort in asking questions in the Workshops, 89% of the students agreed that they are comfortable. Similarly, 89% agreed that they are comfortable with the Workshop leader. By contrast only 38% indicated that they are comfortable asking questions in lecture

Regarding the Workshop activities, students and leaders were in agreement about how time is spent and in general agreement that the current allocation of time to different activities is suitable to their needs. The results of the surveys are presented in Table 4. In columns 1 and 2, the time allocated to different activities is recorded. In columns 3, 4 and 5, the preference of students and leaders for different activities is reported. Both students and leaders reported that more time is devoted to each of two areas than to (c) small-group work: (b) leaders responding to student questions and (e) students presenting solutions. These data suggest that there is more emphasis on "presentation" than on small-group activities. Alternatively, the data may reveal the students' needs for answers and solutions.

**Table 4. Analysis of Workshop Activity**

First, rate each of the following activities according to Workshop time allocated to it. Use the following scale: **1** = little or no time; **2** = some time; **3** = a moderate amount of time; **4** = a great deal of time. Second, indicate whether you would prefer to have **more**, **less**, or **about the same** amount of time for each activity.

| | **Time Allocated** | | **Preference** | | |
|---|---|---|---|---|---|
| | Student %**3,4** Responses | Leader %**3,4** Responses | % Stu (Ldr) **More** | % Stu (Ldr) **Less** | % Stu (Ldr) **Same** |
| a. The Workshop leader presents ideas and methods | 55 | 39 | 32(10) | 4(28) | 64(63) |
| b. The leader responds to student questions | 87 | 73 | 16(15) | 1(15) | 83(70) |
| c. Students work in pairs or small groups | 65 | 54 | 20(35) | 8(5) | 72(60) |
| d. Student work on problems alone | 19 | 15 | 15(10) | 20(28) | 60(63) |
| e. Students present solutions | 77 | 71 | 18(49) | 6(5) | 76(46) |
| g. Hands-on activities such as use of models | 39 | 20 | 45(62) | 6(0) | 51(38) |
| i. Use of technology or computer simulations | 20 | 2 | 40(55) | 12(3) | 48(43) |

The default option for teaching is presentation, of course. It may be that the balance between presentation and small group activities changes as the faculty, leaders, and students gain more experiences with the model. This balance of activities in the Workshop requires much more study. It would be especially interesting to study the relationship of leader training to the balance of activities.

***Faculty involvement.*** A majority of the adapt and adopt faculty members became interested in the project because they were dissatisfied with their traditional teaching/learning methods and environment and because they firmly believe that learning groups help students. They also are hands-on in their approach to teaching and learning; most had already been involved in some form of experimentation. Seven of the twelve professors had previous experience with study groups or some form of group learning: either as part of the lecture or as part of a remedial program or had done research on the subject or had worked with a learning center. These faculty wanted a more organized approach. Two professors were hired explicitly to implement group activities in chemistry education.

When asked, "How do you interact with the professor teaching the Workshop course," student leaders mentioned weekly meetings, e-mail contact, and office hours, and they often commented on the availability and friendliness of the professors. No respondent indicated any incident or situation in which a professor was not available. Clearly, the professors have impressed the student leaders with their support and availability.

Regarding the question of direct participation in the Workshops, faculty reported interesting differences in practice and in theory. Five reported that they do not visit the Workshops, as a matter of principle. Several reasons were given: wanting the Workshop leaders to have full responsibility without anyone appearing to oversee them; this is what was recommended by the Workshop Project. Three professors reported that they occasionally drop in to see how the Workshops are going. One said it is a big mistake to imagine that everything is going well. Three professors reported that they spend the full time in the Workshop; two participate, aiding the student leader, answering questions, and so forth; the third observes.

In spite of the differences in participation in the actual Workshop, the common theme emerging from the interviews is that professors are thoroughly involved with what goes on; none simply turns the program over to a super leader or imagines that the current program will work effectively without ongoing supervision.

***Integral to the course.*** Faculty, Workshop leaders, and students agreed that the Workshop time helps students prepare for tests and achieve higher grades. Professors have generally said that the Workshop time reduces student study time. Students give strong positive endorsements to Workshops in contrast to recitations (when the two were compared in previous studies). The attendance at workshops is high and indicates that students find them to be a worthwhile component of the course.

Regarding the connection between Workshop leaders and involved faculty it is clear that the regular meetings, e-mail exchanges, and individual conversations all strengthen the bonds linking the lecture and Workshop. In addition, professors report that they regularly mention the Workshops and Workshop problems in the lectures, reinforcing the importance of the connection. In a few courses some new material is taught in the Workshops.

***Administrative support.*** Of the faculty members interviewed, eight reported support at the departmental level, although four indicated that this support was not strong or was conditional on success. One reported that the department head did not support the program at all, and one was not sure. Five of the respondents reported strong support from deans; four said they were uncertain of the level of support at his level.

Almost all of those interviewed reported ongoing concerns about funding the program. Funds had been obtained from a variety of sources: NSF, the department, the supplemental instruction center, special funds, and the like.

### Formative Evaluation 2b: Site Visits

Evaluation site visits of one day or more have been made to the University of Rochester, Miami University in Oxford, Ohio, and the University of Kentucky at Lexington, Kentucky. A number of briefer visits have been made to institutions in the New York City area. At these visits, interviews were conducted with Workshop faculty and learning specialists; discussions were held with students and student leaders; Workshops were observed. A focus on the critical role of the leader emerged from these site visits.

***Workshop dynamics.*** The peer leaders generally see themselves as guides and facilitators rather than teachers. At most sites the peer leaders do not have the answers to

Workshop problems but work through them individually and as a group, sometimes under the direction of the professor teaching the course. Students generally accept the fact that answer keys are not forthcoming, although some complain about this early in the course. Leaders are pleased to find that students often use different methods to solve a problem. The peer leaders feel particular responsibility to prepare students for tests, sometimes preparing practice tests and holding special sessions before exams. Workshop leaders vary in whether they seem to take an algorithmic or conceptual approach to learning. Some focus on particular procedures, others on underlying principles. One peer leader said, "I want you to know this and know that you know it so that when it comes up next year, you can do it and not just say 'Oh I remember we did something like this.'"

Workshop leaders were very conscious of the importance of student engagement. Participation always starts slowly but increases during the term. Peer leaders have noticed differences in students' willingness to work with others, the speed at which students can work, and the ability of students to explain something to others. Due in part to the peer leader training and in part to their natural ability to relate to others, the peer leaders give many examples of how they accommodate their approach to student differences.

Students in Workshops have been observed working together, in pairs and small groups, and explaining material to one another. One student noted that "the best way to learn is to teach; you have the opportunity to do that when you're in a study group." They often turn to the peer leader when they are uncertain, and the leader often responds by asking leading questions. "Which is it? Weak or strong? What does that mean?"

Peer leaders have noted that some students think the Workshop can relieve them of study time. While this is marginally true, it is much more to the point that PLTL can make study time much more productive. Peer leaders are generally students with a commitment to academic success and very good study habits. They constantly remind students of the importance of studying the text, completing pre-Workshop assignments, reviewing work, and carefully preparing for exams.

***Coordination of lecture and Workshop***. Professors have noted that PLTL Workshops force a certain organization and pace on the course. When the Workshop coordinator is not the professor teaching the course, the problem is magnified. Similarly, if students do not see a close relationship between the lecture, the Workshop materials, and course grades, they are inclined not to value the Workshops. Tension arises if lectures and Workshops are not well coordinated.

Often students will confide in the Workshop leaders about difficulties with the lecture, their own ability to grasp concepts and procedures, the pace of the course, the materials, and various other aspects of the course. Consequently, the peer leader can become an important conduit between the class and the professor, reporting to the professor about topics or procedures that are not generally understood, and advising the students about the course priorities and direction.

***Problems confronting the Workshop leaders***. In general Workshop leaders are highly enthusiastic about their work. A number have been Workshop or study group leaders several times. At some institutions leaders are active in grooming potential future leaders and take pride in pointing this out.

Nevertheless, the leaders have some problems. The leaders, as undergraduates having taken the course once, are frequently confronted with questions or problems that they cannot instantly answer or solve. Observations and interviews reveal that this potential liability is often turned to an asset. Leaders accepting suggestions from students in the group, or working through a problem with the group, demonstrate that the learning process is not one of memorization and quick recall but of understanding and thinking. Another common problem is simply a reflection of the different personalities, work habits, interests, and attitudes of students. It has been interesting to note that more often than not these differences are resolved with positive learning outcomes for the entire group.

***Benefits to the Workshop leaders***. The Workshop leaders find that explaining material improves their own understanding a great deal. "It's fun for me to be able to break things down to a simpler level, ... to find patterns in what's going on and tell them, 'Look can you find the pattern?'" Another leader reports that "they know I'm going to do something crazy," to introduce a topic or reinforce particular processes. The Workshop leaders report an increase in confidence as a result of their experiences. They also find it satisfying to be able to stand up in front of a group and give an explanation. Many Workshop leaders report that they have developed a serious interest in teaching and hope this will fit with their career plans. The leaders were gratified at expressions of admiration from students. All Workshop leaders said they wished they themselves had taken more courses with Workshops.

## Summative Evaluation: Student Performance

Since the first sites in the project began Workshops more than five years ago, a number of comparison studies have gathered data about the effect of the Workshops on student performance. The results of eight studies are reported here; others are currently underway. As Workshop methods are adapted it will be important to add to the collection of studies, especially to determine the effect of the Workshops in different circumstances and different disciplines.

We have used the percentages of students achieving grades of A, B, and C as the primary comparative measure of student success. The numerator is the number of grades of C– and above; the denominator is the total number of students in the course; that is, the total number of all letter grades plus the number of Incompletes and drops. Various questions can be raised about the samples and about the Workshops themselves in these eight studies, but there is a convergence of evidence indicating that Workshop students earn higher grades than non-Workshop students and also, retention is significantly improved.

***University of Rochester, Professor Kampmeier***. A controlled experiment involved students in the first semester of organic chemistry. Students could opt for the Workshops or the traditional discussion/recitation groups with graduate teaching assistants. About 40% of the students selected Workshops. The same Workshop problems were available to both groups. For students who finished the course, grade point averages were compared for those in the Workshops and for those not participating. Mean GPAs are presented in the following table. The difference is almost half a letter grade and is statistically significant at the 0.05 level.

| U of R: Organic Chemistry | Non-Workshop ($n$ = 171) | Workshop ($n$ = 119) |
|---|---|---|
| GPA | 2.7 | 3.1 |

In order to determine whether there was a self-selecting process by which the more able students selected the Workshops, comparisons were made between Workshop and non-Workshop students' previous GPAs, using the previous semester's GPAs (Spring 1995) and cumulative GPAs. In both cases, there was no significant difference in *prior* GPAs between Workshop and Non-Workshop students as they entered the organic course.

Professor Kampmeier also collected comparative data on percentages of ABC grades for students during recitation years (1992-1995) as compared with Workshop years (1995-1999). The course, text, and tests remained essentially unchanged except for the substitution of

Workshops for recitations. The following table shows the comparison. The difference was statistically significant at the 0.05 level.

| U of R: Organic Chemistry | Non-Workshop (4 sem; $n$ = 1450) | Workshop (4 sem; $n$ = 1554) |
|---|---|---|
| %ABC | 66±1 | 79±5 |

***University of Pittsburgh, Professor Golde.*** Workshop chemistry was first introduced into the general chemistry course at the University of Pittsburgh during the 1995-96 academic year. The course included two and a half hours of lecture, one hour of recitation or Workshop, and a three-hour lab. For Fall 1996, students were randomly assigned to recitation or Workshop groups. Recitation groups consisted of twenty to twenty-four students led by a graduate teaching fellow. Workshop groups consisted of six to eight students led by an undergraduate. The Workshops were provided with problem sheets in which the problems were predominantly nonalgorithmic. Group work was fostered and rewarded. Retention rates were comparable for the two groups. Students took the same exams. The following final grade percentages were achieved.

| Pittsburgh: General Chemistry | Recitation ($n$ = 113) | Workshop ($n$ = 130) |
|---|---|---|
| %A | 20 | 24 |
| %B | 19 | 30 |
| %C | 44 | 36 |
| %ABC | 83 | 90 |

***Queens College, Professor Berkowitz***. At Queens College, Professors William Berkowitz and Schulman taught sections of a general chemistry course using Workshops. In Fall 1995 and Spring 1996 students were randomly assigned to recitation and Workshop groups. Recitation groups were taught by graduate student, Workshops by undergraduate leaders. Workshops began with about twelve students per group, but declined to between five and ten. Recitations began with eighteen to twenty per group and declined to between twelve and eighteen, reflecting the general decline in enrollment as the term progressed. Three course exams and a final exam were given. In each of the six class exams and in both final exams the Workshop students scored higher than the non-Workshop students. Final exam grades are shown in the following table.

| Queens: Final Exam: General Chemistry | Fall 1995 | Spring 1996 |
|---|---|---|
| Recitation Group | 42 ($n$ = 56) | 48 ($n$ = 48) |
| Workshop Group | 51 ($n$ = 57) | 51 ($n$ = 65) |

***St. Xavier University, Chicago, Professor Varma-Nelson***. Professor Varma-Nelson collected grades for two non-Workshop classes (Fall 1993 and Spring 1994) and two Workshop classes (Fall 1994 and Spring 1995) for the same course, Chemistry (General, Organic and Biochemistry) for nursing students. The non-Workshop courses had four hours of lecture. The Workshop course had three hours of lecture and a one-hour Workshop. Professor Varma-Nelson taught all four courses. Data comparing the incoming cohorts are not available. There are significant differences at the 0.05 level for the following results .

| St. Xavier: Gen-Org-Biochem | Non-Workshop ($n$ = 95) 2 semesters | Workshop ($n$ = 116) 2 semesters |
|---|---|---|
| %ABC | 72 | 84 |

*New York City Technical College, Professor Strozak.* Grades were collected for Workshops and for non-Workshop classes of the same course. These classes were taught in Spring and Fall 1995 and Spring 1996. Non-Workshop classes were taught by different professors. Workshop classes were taught by Dr. Strozak. The required Workshops added two hours to a student's total weekly time for the course. The Workshop courses were part of a program for minority students, AMP (Alliance for Minority Participation). A report from the dean at New York City Technical College states that, "It should be noted that because the AMP sections require an additional mandatory two-hour Workshop, the students who sign up for these sections are often those weaker students who tend to sign up at the end of the registration for the only sections remaining."

| NYC Tech: General Chemistry | Non-Workshop ($n$ = 433) | Workshop ($n$ = 131) |
|---|---|---|
| %ABC | 61 | 81 |

*City College.* During the past few years, several members of the chemistry faculty at City College, including Professors Radel, Gosser, Lindsay, Tamargo, and Simms, further developed the sequence of courses for general chemistry. This 103-104 sequence now includes a fully integrated required lab. Workshops have become more standardized with consistent training for all leaders, and a uniform group size. All Workshop leaders now attend a one-credit course on group leadership taught by Ellen Goldstein. Common exams have also been introduced across sections. The following table shows ABC grades for first and second courses in general chemistry. Results in Workshop courses compared with historical averages of ABC grades for non-Workshop courses.

| City College | Non-Workshop | Workshop |
|---|---|---|
| Gen Chem 103.1 %ABC | 38 (historical) | 58 ($n$ = 484; F96, S97) |
| Gen Chem 104.1 %ABC | 52 (historical | 66 ($n$ = 137; S97) |

*The University of Kentucky.* At the University of Kentucky, Professor Joseph W. Wilson offered Workshops for 2%-10% of the students in large general chemistry courses for eight semesters, from Fall 1995 until Spring 1999. Students volunteered to sign up for a one-credit Workshop. On average, composite ACT scores for Workshop students (26.2) were slightly higher than those for non-Workshop students (25.6). Students took the same exams throughout the course. The percentage of Workshop students earning quality grades was considerably higher than that of non-Workshop students, as shown in the following table.

| General Chemistry | Non-Workshop | Workshop |
|---|---|---|
| 1st Semester % ABC | 60±9 | 80±6 |
| 2nd Semester % ABC | 58±4 | 73±11 |

*Conclusions.* The results of each of the eight studies demonstrated positive gains for Workshop students as compared with others, whether non-Workshop classes were conducted with or without recitation. As noted above, these studies vary in the control of other variables. Nevertheless, the Workshop students regularly outperformed the non-Workshop students. This provides a very strong case in favor of the PLTL approach and is a source of confidence that Workshops enhance learning. The success across a wide range of types of courses, types of schools and student bodies indicates that the model is robust and broadly applicable. These studies add to a considerable body of data supporting the value of collaborative learning at the college level.

Additional comparison studies have been initiated and reports will be available on the project Web site as these studies are completed. We invite others to provide new comparisons.

Table 5 summarizes the results of the comparative studies.

**Table 5. Summary of Results of Comparative Studies**

| Institution | Non-Workshop % ABC | Workshop % ABC |
|---|---|---|
| University of Rochester[1] | 66±l ($n$ = 1450) | 79±5 ($n$ = 1554) |
| University of Pittsburgh[2] | 83 ($n$ = 113) | 90 ($n$ = 130) |
| St. Xavier University | 72 ($n$ = 95) | 84 ($n$ = 116) |
| New York Technical College | 62 ($n$ = 433) | 81 ($n$ = 131) |
| City College | 1st semester 38 (historical) | 58 ($n$ = 484) |
| City College | 2nd semester 52 (historical | 66 ($n$ = 137) |
| University of Kentucky | lst semester 60±9 ($n$ = 4554)<br>2nd semester 58±4 ($n$ = 2912) | 80±6 ($n$ = 188)<br>73±11 ($n$ = 151) |

[1]Organic Chemistry.
[2]The organic-biochemistry semester of a General-Organic-Biochemistry course for allied health studies.

**References**


Gillman, L. (1990). "Teaching Programs that Work" *Focus: The Newsletter of the Mathematical Association of America 10*(1): 7-10.

Gokhale, A. "Collaborative Learning Enhances Critical Thinking" http://scholar.lib.vit.edu/ejournals/JTE/jte-v7n1/gokhale.jte-vte-v7n1.html

Halpern, D. F. (1994). *Changing College Classrooms.* San Francisco: Jossey Bass.

Johnson, D. W. and R. Johnson (1989). *Cooperation and Competition: Theory and Research.* Edina, Minn.: Interaction Book Company.

Johnson, D. W., R. Johnson, and K. Smith (1991). *Active Learning: Cooperation in the College Classroom.* Edina, Minn.: Interaction Book Company.

Klemm, W. R. "Using a Formal Collaborative Learning Paradigm for Veterinary Medical Education" http://scholar.lib.vt.edu/ejournal/JVME/V21-1/Klemm.html

National Science Foundation (1995). *Innovating and Evaluating Science Education: NSF Evaluation Forums 1992-1994* (Floraline Stevens, Program Officer). NSF Directorate for Education and Human Resources.

Slavin, R. E. (1983). *Cooperative Learning.* New York: Longman.

Tobias, S. and J. Raphael (1997). *The Hidden Curriculum. Faculty-Made Tests in Science Part 1: Lower Division Courses.* New York: Plenum.

Treisman, U. (1992). "Studying Students Studying Calculus: A Look at the Lives of Minority Mathematics Students in College." *College Mathematics Journal 23*(5): 362-372.

# *Chapter Seven*
# *Vygotsky's Theories of Education: Theory Bases for Peer-Led Team Learning*

**Mark S. Cracolice and Jeffrey A. Trautmann, The University of Montana**

> *"Thought and speech turn out to be the key to the nature of human consciousness."*
> *Vygotsky 1986, 256*

Peer-led team learning works, but *why* does it work? To answer this question, we have studied the works of a psychologist who has yet to gain the stature of the most famous of those whose findings influence education. Nonetheless, we believe that his work is as good as, if not better than, those more well known. Our goal in this chapter is to provide a brief summary of the highlights of his efforts and discuss their applications to peer-led team learning.

Lev Semenovich Vygotsky (1896-1934) is slowly becoming recognized for his important contributions to psychology and education. Originally educated in law and the humanities, Vygotsky became interested in psychology and eventually became one of the preeminent Russian psychologists of his era. His works were suppressed in the former Soviet Union for a number of years, however, and were not known in the west until the 1960s (Van der Veer and Valsiner 1991). Vygotsky's influence on U.S. science and mathematics curricula is therefore still in a relatively early stage.

We begin by addressing Vygotsky's views on concepts, including how they develop and how he classified them. We then look at Vygotsky's views on mental development and how it relates to instruction. This leads to a Vygotskian concept central to understanding peer-led team learning, the *zone of proximal development.* Vygotsky's theories, along with those of other psychologists, fit into the general category of constructivism, so this topic is also addressed, including the heavily Vygotsky-influenced area of social constructivism. Finally, we look at empirical evidence supporting applications of the theories.

## Concept Development

> *"The greatest difficulty of all is the application of a concept, finally grasped and formulated on the abstract level, to new concrete situations that must be viewed in these abstract terms...."*
> *Vygotsky 1986, 142*

In this quote, Vygotsky eloquently summarizes a major hurdle faced by students trying to learn science concepts: applying those concepts. Most science instructors have been frustrated by students who seem to understand the more difficult concepts of the discipline yet fail to make the simpler applications of those concepts. Vygotsky gives us some insight as to why this occurs.

Vygotsky believed that the ability to form concepts proceeds in three phases. The first phase is characterized by trial-and-error thinking and normally is found only in young children. For example, children will group objects based on their location when they first come into the child's visual field, or on temporal relationships such as when the child first notices them.

The second phase, which Vygotsky described as *thinking in complexes*, is characterized by linking objects based on concrete and factual bonds rather than one abstract and logical relationships. This is a lengthy phase of development, characterized by a series of stages that develop in succession. Children first form complexes based on any similar or contrasting relationship between an object that initially attracts their attention and other objects. They progress to grouping objects based on their functional cooperation, such as classifying a shirt, pants, shoes, and tie together into a complex. These functionally based complexes then evolve into chain complexes: a child will group objects based on a consecutive series of relationships; first one, then attention will shift to another, and then to another, and so on. In the next stage, the groupings are based on loosely focused floating and changeable relationships among the objects, such that a grouping may begin with pants but then switch to shorts because they are similar to pants, but then switch to colors because of the color of one pair of shorts, and so on.

The final stage of thinking in complexes is the bridge to the third and final phase of the development of concept-formation. This bridging complex is called the *pseudoconcept* because it resembles a true concept yet is still a complex. This is an important idea for science and mathematics educators. When a student forms a pseudoconcept, he or she will seem to understand the concept when questioned at a level of sophistication that cannot distinguish between concepts and pseudoconcepts; the complex merely coincides with the concept. Vygotsky uses the example of a student who is able to recognize and group triangle-shaped objects without a true understanding of the meaning of triangle. How often have we encountered students who can recognize an example of a concept without really understanding the concept itself?

The third and final phase of concept development occurs when true concepts are based on abstract and logical relationships. A key component of this phase is the ability to pull out the critical elements from the total in which they are rooted. It is also critical to have the ability to form conceptual relationships on the basis of multiple attributes. Vygotsky points out that adults resort to thinking in complexes even after the ability to form true concepts is fully developed. It is certainly a critical issue for science and math educators to design instruction to encourage students to learn how to think at the conceptual level and to avoid curricula that allow students to form complexes that go unchallenged. The Workshop approach, with its built-in questioning of answers, whether right or wrong, is unique in that the peer leader, and possibly other students, challenge and correct Vygotskian complexes, and they help students form scientifically valid concepts.

### Everyday and School-Learned Concepts

> *"We can now reaffirm on a sound basis of data that the absence of a system is the cardinal psychological difference distinguishing [everyday] from [school-learned] concepts."*
>
> *Vygotsky 1986, 205*

Vygotsky distinguished between two different ways that concepts can be acquired, which we will label as *everyday concepts* and *school-learned concepts*.* Everyday concepts are those learned from everyday experiences. School-learned concepts are those learned in school.

A key postulate of Vygotsky's theories is that the development of school-learned concepts runs ahead of the development of everyday concepts. The two types of concepts constantly influence each other and develop in opposite directions, with everyday concepts

---

*Vygotsky used the term *scientific* or *nonspontaneous* concepts to label what we will call *school-learned* concepts. We do this to avoid confusion with other beliefs about scientific concepts. We also use *everyday* concepts in place of Vygotsky's *spontaneous* concepts.

developing toward a more conscious usage, and school-learned concepts developing toward a more concrete, or applied, level.

Vygotsky distinguished clearly between the two types of concepts. They evolve under different conditions; everyday and school-learned concepts develop in different environments and are learned with different attitudes. Everyday and school-learned concepts are also represented differently in the consciousness. In a manner similar to learning a foreign language, a student uses the concepts already acquired (native language) as a mediator between what is already known and the new school-learned concepts that are presented. An additional reason for distinguishing between the two types of concepts is because of the potential value in studying concept formation. Finally, it is important to understand the relationship between instruction and school-learned concepts, which we address in the next section in more detail.

## The Relationship of School Instruction to Mental Development

> *"[Higher mental functions'] composition, genetic structure, and means of action [forms of mediation] - in a word, their whole nature - is social. Even when we turn to mental [internal] processes, their nature remains quasi-social. In their own private sphere, human beings retain the functions of social interaction."* *Vygotsky 1981, 164*

Vygotsky, in collaboration with his student Zhozephina Shif, conducted a series of four investigations of the relationship between instruction and development. From their first series of studies, Vygotsky and Shif concluded that the essential difference between oral speech and writing reflects the difference between the two types of activity, one of which is spontaneous and nonconscious, the other, abstract and conscious. Furthermore, the cognitive functions on which writing is based have not even begun to develop when instruction in writing starts. They must build on barely emerging, immature processes. Finally, as with writing, the requisite functions are immature when instruction in arithmetic, grammar, and natural science begins. From all this they surmised that the development of the cognitive foundations of instruction does not precede instruction but unfolds in a continuous interaction with the contributions of instruction.

In their second series of investigations, Vygotsky and Shif examined the temporal relation between the processes of instruction and the development of the corresponding mental functions. They found that instruction usually precedes development. The student acquires certain skills in a given area before he or she learns to apply them consciously and deliberately. Additionally, instruction has its own sequences and organization; it follows a curriculum and timetable, and it never completely coincides with the developmental process for a given individual. For instance, the third or fourth step of instruction may add little to a student's understanding of arithmetic, but during the fifth step, something may "click." For another student, it might be the fourth or perhaps the seventh step when the transition occurs. Thus, the turning points at which a general principle becomes clear to the student cannot be set in advance by the curriculum. They concluded that when the student learns some operation of arithmetic or some other school-learned concept, the development of that operation or concept has only begun. Furthermore, the path of that development does not coincide with the path of instruction. A critical Vygotskian conclusion is that *instruction precedes mental development*.

In their third series of studies, Vygotsky and Shif focused on transfer of training among higher functions that could be expected to be meaningfully related. Their results indicated that intellectual development is not compartmentalized according to topics of instruction. Instruction in a given subject influences the development of higher functions far beyond the confines of that particular subject. Consequently, all the basic school subjects act as a formal discipline, each facilitating the learning of the others.

Finally, in the fourth series of investigations, Vygotsky and Shif developed a new method of assessing a student's level of mental development at a particular time. Before Vygotsky, most psychological investigations of school learning measured the level of mental development by making the students solve certain standardized problems. The problems that they were able to solve by themselves supposedly indicated their mental age, but Vygotsky postulated that the tests measured only the completed part of development. Their approach was to pair students of the same mental age and give each of them more difficult problems than they could manage on their own. They gave them slight assistance as necessary. In a typical result, one student could solve problems designed for students four years beyond his or her standard mental age, whereas the other could go only one year beyond his or her standard mental age. The discrepancy between the standard mental age and the mental age indicated when solving problems with assistance is what Vygotsky called the *zone of proximal development*.

**The Zone of Proximal Development**

> *"...developmental processes do not coincide with learning processes. Rather, the developmental process lags behind the learning process; this sequence then results in zones of proximal development."*
> *Vygotsky 1978, 90*

Let us begin with Vygotsky's (1978) definition of the zone of proximal development "It is the distance between the actual developmental level as determined by independent problem solving and the level of potential development as determined through problem solving under adult guidance or in collaboration with more capable peers." We believe that this concept is central to the theoretical understanding of peer-led team learning.

One important application of the concept of the zone of proximal development is in understanding the role of imitation in learning. Students can perform intelligent, conscious imitation only within their zones of proximal development. By imitating, students can carry out actions beyond their capabilities as defined by the lower end of their zone. Vygotsky believed that imitation is indispensable in acquiring school-learned concepts because the process leads the student to new, higher developmental levels.

Because of the ability of humans to imitate, Vygotsky said, "human learning presupposes a specific social nature and a process by which children grow into the intellectual life of those around them." This led him to hypothesize that the developmental process follows, or lags, the learning process. This is what causes zones of proximal development: the initial learning of one concept provides the basis for learning more complex concepts. Additionally, even though development and learning are directly related, they are neither equal nor parallel. They interact in a complex, dynamic manner.

**Implications of Vygotsky's Theories for Instruction**

> *"Learning science should involve the gradual integration of personal experience and knowledge into the complex systems of models and theories, and the ways of thinking, that scientists use to explain natural phenomena. This makes teaching a little messier than some would like it to be; one cannot teach an important concept, assume that it has been learned, and never return to it."*
> *Howe 1996, 47*

Howe (1996) outlines four applications of Vygotsky's theories that should be applied to the design of instruction in science courses. First, utilization of manipulative objects, such as often occurs in lab or when using models to understand particulate-level concepts, is necessary but not sufficient. Social interaction, in which students use language as a tool to reflect on their experiences and discuss how their experiences fit into a larger system, is also essential. Second, context is more important than cognitive demand. Third, the sociocultural aspects of the classroom must be designed to facilitate instruction. Students gain knowledge through interpersonal interactions, and thus the nature of the relationships among students and instructor has a direct influence on the learning process. Finally, the role of language must be explicitly incorporated in the curriculum design. Because students use language as a tool in thinking and communicating, they must learn the relationships between their everyday language and the school-learned language used by scientists.

We believe that the Workshop model of instruction incorporates all of Vygotsky's theories into science and mathematics instruction. That is why peer-led team learning works. Peer leaders embody Vygotsky's postulate that instruction precedes development. The peer leader is a guide to development. Because peer leaders have completed the course, they have a natural tendency to encourage and nurture the students to develop to their level. This takes place in a rich social environment in which misconceptions are challenged, and language consistent with that used by scientists is applied to newly formed concepts. With problems and materials that are situated within the students' zones of proximal development, the PLTL model provides the mechanism by which students can construct new conceptual understanding. The requirement to verbalize what they already know and discuss how that relates to the newly acquired information and ideas provides a unique learning opportunity that is often omitted in standard curriculum designs.

Peer leaders are also critical to the success of the Workshop model because they provide a learning structure known as *scaffolding*. An example of scaffolding is a parent guiding a child in taking his or her first steps. A great deal of support is provided at first, but this support is gradually removed as the child shows an increased ability to walk independently. Technically, scaffolding is an application of Vygotsky's assisted learning concept. He believed that the tools necessary to perform higher mental functions are initially external, but through the assistance of a mediator, they become internalized, and then they are available to be used in higher-order thinking. In peer-led team learning, the peer leader and the other students on the team are the mediators who provide the scaffolding to guide one another to learn the tools and skills that are needed to function at higher levels.

### Constructivism

> *"No idea has more implications for teaching and learning than the realization that knowledge cannot be transmitted intact from one person to another. No matter how I try, I can never transfer an idea from my head to yours."* *Herron 1996, 17*

Vygotsky's theories, and those of many other psychologists, most notably Piaget's, fall under the umbrella of constructivism. The essence of the constructivist theories is that cognitive development is a process by which people convert information into understanding. Based on their previous experiences and knowledge, the context in which the new information is presented, and their procedural knowledge students construct their own personal meaning out of the new information. Instructors cannot simply give their understanding of concepts to their students; students will necessarily construct their own meanings based a complex array of factors unique to their own personal experiences and development.

One of the most widely read papers on constructivism in the chemical education community was published by Bodner in 1986. He provided a simple, straightforward definition: "Knowledge is constructed in the mind of the learner." Bodner further refined the constructivist model by identifying a number of implications for effective teaching, starting with the suggestion that the instructor needs to change from teacher to learning facilitator. This implies that an instructor needs to establish a dialogue between students and instructor and find ways to keep students actively involved in the learning process. This certainly is accomplished in a Workshop setting.

One application of constructivism is the learning cycle instructional strategy (see Lawson, Abraham, and Renner 1989 for a comprehensive account). This three-part strategy for teaching begins with the *exploration* phase. Here, students explore new observations, usually with the objective of finding a pattern in the data or recognizing a conflict between the observations and their understanding. The second phase, called *term introduction*, is where scientific vocabulary is introduced to give students a way to discuss the new phenomenon. The third phase, *concept application*, is where students apply the newly learned concept to other situations. Much of the curriculum developed for PLTL follows a learning cycle model.

Another example of an application to instruction that is derived from constructivism is the *generative learning model* (Cosgrove and Osborne 1985; Osborne and Freyberg 1985; Wittrock 1986). The general idea of this model is to help students learn strategies to deal with new information in a way that generates learning. In order to do this, these authors propose a four-phase model of instruction. First, the instructors must assess students' prior knowledge that relates to the concept to be taught. Second, the instructors must assess their own views about the concept, the views of their students, and those of other scientists. Third, the instructors must allow the students to explore the concept, most preferably in a real context. Finally, the instructors must provide an opportunity for the students to consolidate and elaborate their ideas about the concept must be provided. Well-designed materials and a well-trained peer leader will guide the Workshop student through these steps.

**Social Constructivism**

> *"The scientific disciplines are socially constructed; thus, classroom activity which introduces students to various scientific ideas should be fashioned in similar ways."*
>
> *Atwater 1996, 825*

A definition of social constructivism is provided by Oldfather et al. (1999): "Theory of knowledge that holds that knowledge is constructed within a social context through language and other sign systems. A social constructivist perspective focuses on learning as sense-making and not on the acquisition of knowledge that 'exists' somewhere outside the learner." Vygotsky can certainly be called a social constructivist. His views can be contrasted with those of Piaget, who believed that the learner primarily built concepts through interaction with the environment. Vygotsky, on the other hand, believed that there was an essential role for language and communication with an instructor or peers while interacting with the environment.

Oldfather et al. (1999) outline a number of characteristics of a social constructivist classroom (p. 74). It is interesting to use these characteristics to compare a typical peer-led team learning Workshop with traditional classroom.

- A primary goal of the classroom is collaborative construction meaning. The focus is on sense-making rather than on construction of a single right answer. Errors are viewed as a natural part of learning and considered to be opportunities for growth. Teachers and students search for meaningful connections between what they know and what they are learning. Everyone shares the ownership of knowing. The teacher is not the sole authority for knowledge.
- Teachers pay close attention to students' perspectives, logic, and feelings.
- The teacher is teaching and learning. Students are teaching and learning. Everyone is asking questions and pursuing them.

- Teaching and learning are based on social interaction. The talk is both structured and unstructured. The flow of ideas and information is multidirectional.
- Everyone is treated as a whole person. Students' physical, emotional, and psychological needs are considered along with their intellectual needs.
- The teacher and students believe that everyone can succeed. Assessment is based on each individual's progression and not exclusively on competitive norms.

**Research on Cooperative Learning**

> *"Small-group learning is clearly successful in a great variety of forms and settings and holds considerable promise for improving undergraduate SMET [science, mathematics, engineering, and technology] education."* *Springer, Stanne, and Donovan 1999, 43*

Educational research provides strong evidence that cooperative learning is an effective tool in helping students learn scientific concepts. One of the most convincing studies, that of Springer, Stanne, and Donovan (1999), utilized a meta-analysis method that cumulates and integrates the results of all of the research conducted in a specified area. This analysis looked at 39 studies of cooperative or collaborative learning among two to ten students involving undergraduates in science, mathematics, engineering, or technology courses in North American institutions. The findings showed that students who worked in small groups showed greater achievement, persisted through the courses or programs to a greater extent, and had more favorable attitudes than control-group students. Moreover, the size of the effect was relatively large. A student who would have scored in the 50th percentile on a standardized achievement test without small-group learning would move to the 70th percentile with small-group learning. The analysis of persistence showed that small-group learning reduces attrition from courses and programs by 22%. The impact on students' activities was almost twice the average effect for other classroom-based educational interventions.

Other large studies provide additional evidence in support of the effectiveness of cooperative learning. Johnson and Johnson (1992) conducted a meta-analysis of 323 studies of cooperative learning and reported that it was significantly more effective in enhancing learning than either competitive or individualistic learning. Slavin (1990, 1991) reported that the critical conditions for successful cooperative learning are that groups must be sure that all their members are learning, and that those groups that do so must receive appropriate rewards.

Bianchini (1997) studied the conditions necessary for fostering quality discourse in group-learning settings and found four important components. First, group tasks must be coherent in procedure and purpose, with an emphasis on making connections between the topic immediately at hand and other topics in the course. These connections must be made explicit to students. Second, instructors must clearly specify the purpose, content, and processes to be learned as a result of the group activity. This benefits the students and the instructor, who will more clearly understand the necessary prerequisites, the potential pitfalls, and the appropriate questions. A third condition is an awareness of assumptions of prior knowledge required by the task. The constructivist model explicitly acknowledges that what a student knows at the start influences the way that instruction will affect their conceptions. Finally, Bianchini states that students need to broaden their definitions of science, with an emphasis on making connections among scientific concepts and disciplines and between science and their everyday lives. They must learn to view science as a powerful way of thinking about their world and not as just an esoteric game played by a select few.

## Conclusions

> *"There is a need for an educational theory base for science teaching. That theoretical basis for science teaching could be based upon educational purpose, a developmental model of learning and the structure of the discipline of science."* *Renner and Marek 1990, 245-246*

The data about peer-led team learning show the effectiveness of the model. Vygotsky's research into the relationship between thought and language allow us to look into the data. In particular, his writings on the social nature of cognitive development give us a foundation from which we can understand the power of the PLTL. Vygotsky insists that social interaction with those more capable (knowledge developed) is essential to resolve the disequilibrium that occurs when what is thought to be known collides with new observations about nature. Language is the medium for that social interaction. The locus for that interaction is the students' zone of proximal development. Workshops are designed to fit the zones of proximal development of the students. The challenge is appropriate, and the leader and the students provide the stimulus and the models for further development by the individual. *Instruction precedes development.*

All proponents of the constructivist model, including Vygotsky, acknowledge that what students know before they are instructed is critical to their understanding of the new concept. A rich body of research in science education identified many of the alternative conceptions that can occur when new data interact with old thoughts. Some results suggest that building alternative conceptions is a natural process of learning that occurs before valid scientific conceptions develop. In peer-led team learning, we give students a chance to consider, evaluate, and challenge the alternative conceptions that develop as learning occurs. Classic instructional strategies do not allow time or opportunity to consider what a student thinks, even though this is central to the learning process. The Workshop provides both time and opportunity.

## References

Atwater, M. M. (1996). Social Constructivism: Infusion into the Multicultural Science Education Research Agenda. *Journal of Research in Science Teaching 33*(8): 821-837.

Bianchini, J. A. (1997). Where Knowledge Construction, Equity, and Context Intersect: Student Learning of Science in Small Groups. *Journal of Research in Science Teaching 34*(10): 1039-1065.

Bodner, G. M. (1986). Constructivism: A Theory of Knowledge. *Journal of Chemical Education 63*(10): 873-878.

Cosgrove, M. and R. Osborne (1985). Lesson Frameworks for Changing Children's Ideas. In Osborne, R. and P. Fryberg, Eds., *Learning in Science: The Implications of Children's Science*. Auckland, New Zealand: Heinemann, 101-111.

Herron, J. D. (1996). *The Chemistry Classroom: Formulas for Successful Teaching.* Washington, D.C.: American Chemical Society.

Howe, A. C. (1996). Development of Science Concepts Within a Vygotskian Framework. *Science Education 80*(1): 35-51.

Johnson, D. and R. Johnson (1992). In Sharon, S., Ed., *Cooperative Learning: Theory and Research*. New York: Praeger, 23-37.

Lawson, A. E., M. R. Abraham, and J. W. Renner (1989). *A Theory of Instruction: Using the Learning Cycle to Teach Science Concepts and Thinking Skills.* National Association for Research in Science Teaching: NARST Monograph No. 1.

Oldfather, P., J. West, with J. White and J. Wilmarth (1999). *Learning through Children's Eyes: Social Constructivism and the Desire to Learn.* Washington, D.C.: American Psychological Association.

Osborne, R. and P. Freyberg (1985). *Learning in Science: The Implications of Children's Science.* Auckland, New Zealand: Heinemann.

Renner, J. W. and E. A. Marek (1990). An Educational Theory Base for Science Teaching. *Journal of Research in Science Teaching 27*(3): 241-246.

Slavin, R. E. (1990). *Cooperative Learning: Theory, Research, and Practice.* Englewood Cliffs, N.J.: Prentice-Hall.

Slavin, R. E. (1991). Synthesis of Research on Cooperative Learning. *Educational Leadership 48*(5): 71–82.

Springer, L., M. E. Stanne, and S. S. Donovan (1999). Effects of Small-Group Learning on Undergraduates in Science, Mathematics, Engineering, and Technology: A Meta-Analysis. *Review of Educational Research 69*(1): 21-51.

Van der Veer, R. and J. Valsiner (1991). *Understanding Vygotsky: A Quest for Synthesis.* Cambridge, Mass.: Blackwell.

Vygotsky, L. S. (1978). *Mind in Society: The Development of Higher Psychological Processes*, Cole, M., V. John-Steiner, S. Scribner, and E. Souberman, Eds. Cambridge: Harvard University Press.

Vygotsky, L. S. (1981). The Genesis of Higher Mental Functions. In Wertsch, J. V., Ed., *The Concept of Activity in Soviet Psychology.* Armonk, N.Y.: Sharpe.

Vygotsky, (L. 1986). *Thought and Language*, Kozulin, A., Trans. and Ed. Cambridge: The MIT Press.

Wittrock, M. C. (1986). Students' Thought Processes. In *Handbook of Research on Teaching*, Wittrock, M. C., Ed., 3rd ed. New York: Macmillan.

# Chapter Eight
# An Introduction to Theory and Research on Promoting Student Motivation and Autonomous Learning in College-Level Science

**Aaron E. Black*, University of Rochester**

The purpose of this chapter is to introduce aspects of psychological theory and research on student motivation in the classroom and to share the findings of research evaluating those characteristics of the Workshop Chemistry Project. The most critical component of Workshop Chemistry is the use of peer-led study groups. These groups are meant to increase student engagement with the material, promote active learning, decrease interstudent competition, enhance student problem-solving skills, provide social support, and reduce anxiety. Such goals are attempts to make instruction in college-level chemistry more student-centered and to move away from a traditional "sink or swim" atmosphere. Workshop Chemistry is predicated, in part, on the notion that reducing the "cutthroat" atmosphere in college-level chemistry can remove barriers to student success. This chapter is about the converse: identifying the positive aspects of the Workshop environment that actively promote student engagement and learning.

The effectiveness of the peer leaders in supporting student autonomy in the learning process is particularly interesting. For example, we know from the education literature that students tend to learn more and have greater conceptual grasp of the material when they are interested in and engaged by the learning process (Grolnick and Ryan 1987, 1989; Williams and Deci 1996). Students engaged in autonomous learning take maximum ownership for every step of the process, from the questions they ask in class and preparation for exams to the fundamental ways in which they interact with their classmates and instructors. They are easy to spot by the obvious interest and energy they display. Interest and engagement are both the "fuel" and the results of autonomous learning. Educational environments that facilitate student interest and engagement, by definition, support autonomous learning.

Not all academic environments support student autonomy, however. In the Workshop model, our research has shown that peer leaders are not perceived by their students to be equally supportive of student autonomy (e.g., allowing students to learn in their own ways or equally controlling (e.g., acting like there is only one way to do a problem). The concepts of "autonomy support" and "autonomous learning" as applied to education are reviewed below. *Self-determination theory* (SDT) is the overarching theory and research from which these concepts are drawn.

## Self-Determination Theory and Related Research

Self-determination theory (Deci and Ryan 1985; 1991) views the individual as innately oriented toward the mastery of optimally challenging experiences and understands development

*Correspondence about this chapter may be directed to Aaron Black, Ph.D., Clinical Assistant Professor of Psychiatry, 1357 Monroe Avenue, Rochester NY 14618, or aaron_black80@hotmail.com.

as a dialectical process between the active, evolving individual and forces or influences outside the individual. Intrinsic motivation (reading for pleasure is a common example) is central to this developmental process. Psychological development is thought to be organized around the needs for autonomy, competence, and relatedness. *Autonomy* is the need to feel like the "author" of one's behavior; the person feels a sense of choice about engaging in the behavior and a sense of ownership over the behavior. *Competence* is the need to feel effective in executing tasks that are of some personal importance. Finally, *relatedness* is simply the sense that one is emotionally cared about and close to others. Healthy development in any setting depends on the degree to which the social environment supports or thwarts the satisfaction of these basic needs.

SDT suggests that behaving for reasons that are autonomous (e.g., out of interest) versus controlled (e.g., out of compulsion) predict a wide range of subjective and objective indices of educational achievement, motivation, psychological development, and personal well-being. Self-determination theory suggests that educational environments that support autonomy are likely to shift students' reasons for studying from more controlled to more autonomous. Students who study for more autonomous reasons tend to enjoy learning more and achieve better educational outcomes than students who study for less autonomous reasons. These latter students tend to feel greater anxiety and pressure and enjoy learning less. A great deal of past research provides support for these concepts and suggests ways in which the central tenets of self determination theory can be applied to enhance student instruction and learning in the classroom (e.g., Deci et al. 1994; Grolnick and Ryan 1987, 1989; Miserandino 1996; Williams et al. 1996; Williams and Deci 1996; Ryan and Connell 1989; Ryan, Plant, and O'Malley,1995).

The concept of autonomy support (Deci and Ryan 1985; Williams, Deci, and Ryan 1995) as described by SDT involves a set of behaviors whereby one person actively supports the autonomy needs of another person. In hierarchical relationships such as instructor-student relationships, autonomy support describes how the authority figure takes the other's perspective, acknowledges the other's feelings and point of view, and provides the other with pertinent information and opportunities for choice while minimizing the use of pressure or control. Several examples of student ratings of leader autonomy support are given next.

**Examples of Peer-Leader Autonomy Support in Workshop Chemistry**

I feel that my PL provides me with choices and options.
I am able to be open with my PL during group meetings.
I feel understood by my PL.
I feel that my PL accepts me.
My PL makes sure that I really understand the goals of the course and what I need to do.
My PL conveys confidence in my ability to learn chemistry.
My PL encourages me to ask questions.
I feel a lot of trust in my PL.
My PL handles people's emotions very well.
My PL listens to how I would like to do things.
My PL answers my questions fully and carefully.
My PL tries to understand how I see things before suggesting a new way to do things.
I feel that my PL cares about me as a person.
I feel good about how my PL talks to me.
I am able to share my feelings with my PL.

One way to understand the positive effects of autonomy support is in terms of self-esteem. All learning involves the mastery of previously unknown skills and concepts. Being able to say "I don't understand" is central to the learning process, but it requires the student to be able to tolerate the challenge to self-esteem inherent in an admission of ignorance (i.e., feeling "stupid"). Autonomy support tends to create an atmosphere in which students feel more open to acknowledging confusion, without the fear that they will be embarrassed. Having a person in a position of authority recognize and validate their point of view is central to that shift in focus. When students worry less about protecting their self-esteem, they are much freer to pay full attention to the task at hand and enjoy the intrinsic satisfaction that comes with mastering new knowledge and skills. The result is a deeper sense of engagement in the learning process and better educational outcomes.

When the SDT perspective is applied to student learning in introductory college science courses, one would expect autonomy support from the PL to predict greater mastery of course material, positive educational values (an emphasis on learning as opposed to grades), and positive subjective experiences in the course. Thus, the research sought to examine the effects of students' initial levels of autonomy and the students' perceptions of the degree of autonomy support provided by their peer leaders on a range of subjective experiences and grades in the course.

**Brief Overview of Study Methods**

The study was conducted using participants drawn from an introduction to organic chemistry course at the University of Rochester during Fall 1996. The course itself was organized around the teaching and curricular goals of the Workshop Chemistry Project (Gosser et al. 1996). The peer leaders were trained to facilitate group problem-solving processes. Participation in the 42 Workshops was expected, and attendance rates were high (mean = 11.8, out of 13 meetings, S.D. = 1.8). Students were randomly assigned to Workshop groups. The students ($n$ = 116) completed the questionnaires during two different lecture periods: two days prior to the first exam and several days after the fourth exam, but before the final exam.

**Study Findings**

Several significant findings emerged from the study (Black and Deci 1998). First, the *initial* autonomy of students' reasons for studying organic chemistry was consistently related to their *final* subjective experience of the course. After controlling statistically for the effects of course grades, students with more autonomous motivations at the start perceived themselves as more competent at chemistry, felt greater interest and enjoyment, and were less anxious and grade-oriented at the end than students with lower starting autonomy. Second, the leader matters. After controlling for exam performance, more positive student ratings of their Workshop leaders' support of their autonomy were linked to a greater sense of competence, more interest and enjoyment, and less course-related anxiety. Finally, the leaders can effect change. The leaders' support of the autonomy of their students led to increased autonomy in the students' reasons for participating in the course; i.e., over the course of the term, students who perceived their group leader as demonstrating greater support for their autonomy became more likely to study chemistry for autonomous reasons, such as out of interest or enjoyment.

We also found that support of student's autonomy did more than just improve the students' subjective experience of the course; for some students, it *also* led to enhanced academic performance. For students taking the course for more autonomous reasons, autonomy support was uncorrelated with final grades. Presumably, these students were already sufficiently motivated and engaged that instructor autonomy support by the leader contributed little to enhance their performance. In contrast, for students taking the course for more controlled reasons, the leaders' support of their autonomy was strongly predictive of final grades. The magnitude of the effect of autonomy support on grades was equal to the effect for cumulative grade point average prior to taking the course; i.e., *autonomy support by the leader was as important as cumulative GPA in predicting final grades* . The implication of this finding is that the students who are less interested in the course material at the start are the students whose final grades depend most on the leaders' support for their autonomy.

## Additional Research on Workshop Chemistry

In addition to replicating the original study findings, a more recent study (Church 1998) examined the classroom environment in both the lectures and Workshops. The goals were to understand how the social context influences students to adopt certain goals, and the effects of those goals on performance. Specifically, three types of goals were examined: *performance-approach* goals, *performance-avoidance* goals, and *mastery* goals (Elliot and Harackiewicz 1996). Performance-approach goals are those in which students are focused on demonstrating their competence relative to others; for example, "it is important for me to do better than others". In contrast, performance-avoidance goals are focused on avoiding unfavorable judgments about one's competence; for example, "My fear of performing poorly in this class is often what motivates me." Finally, mastery goals are focused on the attainment of a sense of competence by becoming effective at, the subject matter; for example, "I want to learn as much as I can from this class." The study examined the extent to which the following experiences in the classroom affect the goals that students adopt: how engaging the tasks are, whether students are made to feel embarrassed for mistakes, the difficulty of the course material, how much public focus in placed on student evaluation, how student performance is recognized interpersonally, and whether instructors have negative or positive expectations for student performance.

A number of important findings emerged from the relationships between the interpersonal atmosphere in the classroom and the types of goals that students adopt. Students tended to adopt performance-approach goals when they were engaged by the Workshops, when they viewed their professor and Workshop leader as supporting their autonomy, and when they did feel that their performance was openly evaluated in the lecture or Workshop. Students tended to adopt performance-avoidance goals when they did feel that their performance was openly evaluated in the lecture or Workshop. Engagement in the lecture and Workshop and professor and group leader autonomy support were unrelated to the adoption of performance-avoidance goals. Students tended to adopt mastery goals when they were engaged by the lectures and Workshops, when they viewed their professor and Workshop leader as supporting their autonomy, and when they did not feel that their performance was openly evaluated in the lecture or Workshop. Thus, the classroom environment was shown to have a strong effect on the types of goals that students adopted in Workshop Chemistry.

What then are the effects of the different types of goals on performance? Students who adopted performance-approach goals experienced a greater likelihood of continuing in the subject area and achieved a higher grade in the course; however, performance-approach goals were not related to interest and enjoyment of the course or attendance. Students who adopted performance-avoidance goals experienced decreased interest and enjoyment of the course, had a decreased likelihood of continuing in the subject area, and achieved a lower grade in the course. Performance-avoidance goals were not related to attendance. Students who adopted mastery goals experienced increased interest and enjoyment of the course, had a greater likelihood of continuing in the subject area, missed fewer lectures, and achieved a higher grade in the course.

In summary, the Workshop Project shows a great deal of promise in redefining instructional methods in college-level science. Collectively, the research findings (Black and Deci 1999; Church 1998) suggest that traditional models of introductory science instruction, with emphases on large lectures, limited interpersonal interactions and support, and a "sink-or-swim" atmosphere, may undermine students' interest and performance by failing to provide contextual support for student needs that are central to autonomous learning. In contrast, the structure of the Workshop model seems to be a good mechanism for supporting the autonomy of individual students. The studies reported here show that strong autonomy support by the Workshop leader predicts enhanced subjective experiences for all students and improved performance for students who take the organic chemistry for controlled reasons (e.g., it is a required course). A closer examination shows that the environment in the lectures and Workshops directly influences the type of goals students adopt and that these goals affect the outcomes. Strong autonomy support correlates with performance-approach and mastery goals, and these goals lead to higher performance and a desire to continue in the subject area.

**References**


Black, A. E. and E. L. Deci (1999). The Effects of Instructor Autonomy Support and Student Self Regulation on College-Level Learning: A Self-Determination Theory Perspective. *Science Education*, in press.

Church, M. A. (1998). Classroom Context and Achievement Behavior: The Role of Contextual Factors in the Adoption of Approach and Avoidance Achievement Goals. Unpublished doctoral dissertation, University of Rochester.

Deci, E. L., H. Eghrari, B. C. Patrick, and D. R. Leone (1994). Facilitating Internalization: The Self Determination Theory Perspective. *Journal of Personality 62*: 119-142.

Deci, E.L. and R. M. Ryan (1985). *Intrinsic Motivation and Self Determination in Human Behavior.* New York: Plenum.

Deci, E. L. and R. M. Ryan (1991). A Motivational Approach to the Self: Integration in Personality. In Dienstbier, R., Ed., Nebraska Symposium on Motivation: Vol. 38. *Perspectives on Motivation*. Lincoln: University of Nebraska Press.

Elliot, A. J. and J. M. Harackiewicz (1996). Approach and Avoidance Achievement Goals and Intrinsic Motivation: A Mediational Analysis. *Journal of Personality and Social Psychology 70*: 461-475.

Gosser, D., V. Roth, L. Gafney, J. Kampmeier, V., Strozak, P. Varma-Nelson, S. Radel, and M. Weiner (1996). Workshop Chemistry: Overcoming the Barriers to Student Success. *The Chemical Educator* 1 [on-line serial]. Available at URL:http://journals.springer-ny.com.chedr.

Grolnick, W. S. and R. M. Ryan (1987). Autonomy in Children's Learning: An Experimental and Individual Difference Investigation. *Journal of Personality and Social Psychology 52*: 977-1077.

Grolnick, W. S. and R. M. Ryan (1989). Parent Styles Associated with Children's Self-Regulation and Competence in School. *Journal of Educational Psychology 81*: 143-154.

Miserandino, M. (1996). Children Who Do Well in School: Individual Differences in Perceived Competence and Autonomy in Above-Average Children. *Journal of Educational Psychology 88*: 203-214.

Ryan, R. M. and J. P. Connell (1989). Perceived Locus of Causality and Internalization: Examining Reasons for Acting in Two Domains. *Journal of Personality and Social Psychology 57*: 749-761.

Ryan, R.M., R. W. Plant, and S. O'Malley (1995). Initial Motivations for Alcohol Treatment: Relations with Patient Characteristics, Treatment Involvement, and Dropout. *Addictive Behaviors 20*: 279-297.

Williams, G. C. and E. L. Deci (1996). Internalization of Biopsychosocial Values by Medical Students: A Test of Self Determination Theory. *Journal of Personality and Social Psychology 70*: 767-779.

Williams, G. C., E. L. Deci, and R. M. Ryan (in press). Building Health-Care Partnerships by Supporting Autonomy: Promoting Maintained Behavior Change and Positive Health Outcomes. In Hinton-Walker, P., Suchman, A. L., and R. Botehlo, Eds., *Partnerships, Power, and Process: Transforming Health Care Delivery*. Rochester, N.Y.: University of Rochester Press.

Williams, G. C., V. M. Grow, Z. Freedman, R. M. Ryan, and E. L. Deci (1996). Motivational Predictors of Weight Loss and Weight Loss Maintenance. *Journal of Personality and Social Psychology 70*: 115-126.

# *Chapter Nine*
# *The View from Industry*

**MacCrae Maxfield, Allied Signal**

## Assessing the Importance of Workshops in Undergraduate Chemistry Education

Many of the graduates of undergraduate chemistry education seek employment in industries that manufacture chemical-based products, for example, polymers, petroleum products, pharmaceuticals, and the like. Managers and senior technical personnel in these industries see themselves as *customers* and chemistry departments as *suppliers* of employees trained in chemistry. Successful suppliers always ask their customers what they want. We asked a large sample of managers and senior technical personnel what they are looking for in applicants for jobs that require chemistry training, and whether their criteria would be better met if Workshops were used in undergraduate chemistry education.

To gather data, we designed and distributed a questionnaire that described the Workshop approach in chemistry courses and included a list of twenty-seven attributes that are desirable for employees whose jobs require college-level chemistry education. The selected attributes are based on tasks that personnel are expected to undertake at Allied Signal and at other corporations having diverse manufacturing interests. The twenty-seven attributes can be organized into the following five categories:

- Knowledge of sciences
- Experimental skills
- Data gathering and handling
- Problem-solving skills
- Communication and team skills

The questionnaire asked employers to rate both the importance of the twenty-seven attributes and how well current undergraduate chemistry develops these attributes. Employers were also asked to add and rate additional attributes and to indicate whether they were interested in learning more about chemistry Workshops.

The questionnaire was sent to researchers, research managers, product managers, quality assurance managers, technology managers, and vice presidents from fifteen major companies. These individuals were chosen because they participate in the evaluation of job applicants and performance reviews of coworkers.

The initial distribution of sixty explanatory letters and questionnaires elicited 25 responses. The respondents rated the importance of the attributes on a scale of 0 to 4, where 0 signified no importance and 4 signified high importance. They also used a scale of 2 to 0 to rate how current chemistry education *prepares* students with respect to these same attributes. On this scale, 2 signified no preparation and 0 signified very good preparation. The product of these two scores, *preparation* × *importance*, is a measure of the opportunity for improvement in undergraduate education.

***Summary of questionnaire responses.*** The respondents indicated that students are best prepared in basic science concepts, mathematics, computer skills, ability to perform experiments, and all aspects of chemistry but receive limited preparation in developing skills in related fields (e.g., biology, chemical engineering, and environmental science), problem-solving, communication and team skills, and relating practical applications to basic science concepts.

Respondents rated the attributes in the communication and team-skills category as six of the ten most important attributes for chemistry-trained employees in companies today. Communication with others was judged most important (3.8/4.0). The ability to relate applications to scientific principles was tied with working as part of a team for second place (3.6/4.0); both were more important than the ability to perform experiments (3.5/4.0) and understanding basic science concepts (3.3/4.0). Finally, the ability to define a logical approach to abstract problems (3.4/4.0) was considered important, but working with others (3.5/4.0) was considered as important.

**Opportunities to Improve Undergraduate Chemistry Education**

The questionnaire rating system allowed the manufacturing industries to highlight opportunities for improvement in undergraduate chemistry education. When the product of the *importance* and *preparation* ratings for each attribute is less than four, the current curriculum is doing a good job, or that attribute is of little relative importance to the employer. For example, the attribute "understanding basic science concepts" is regarded as very important (3.3) but well treated in current undergraduate chemistry (0.6); the *importance-preparation* product is about 2, and there is no need for improvement. On the other hand, when *importance* is rated high and *preparation* is rated poor, the *importance-preparation* product is 4 or higher and directs our attention to the **ten top opportunities for improvement in undergraduate chemistry education.** (The *importance-preparation* product is given in parentheses.)

**Communicate with others (5.0)**
**Work as part of a team (4.7)**
**Work with others, different jobs (4.6)**
**Report writing (4.1)**
**Statistical methods (4.0)**

**Oral presentation (4.9)**
**Relate applications to principles (4.6)**
**Approach to abstract problems (4.4)**
**Work with others, same job (4.1)**
**Complex problem solving (4.0)**

Analysis of questionnaire responses indicates that the industrial employers would like chemistry-trained employees whose education includes greater preparation in communication, team skills, in relating applications to scientific principles, and in problem solving, without sacrificing thorough preparation in basic science concepts and experimental skills.

***The importance of team skills.*** Team skills, as understood by industrial researchers, managers and, indeed, virtually all employees of major corporations, refer to the skills necessary to reach decisions in a short time on issues of interest to several people and then to act on those decisions. Team skills include reporting and analyzing information, debating, persuading, sorting fact from opinion, prioritizing, gathering resources, and recording. Teamwork does not mean going along with the crowd but rather bringing knowledge and powers of logic and persuasion to bear on a problem to assure that the best solution is selected by the team and then acting on the decision to assure an impact. More and more companies are adopting policies that encourage employees to initiate and participate in teams to grapple with problems facing the company.

***The potential impact of Workshops.*** Industrial teams frequently use a decision matrix to sort out multiple outcomes and behaviors associated with a problem. In the decision matrix, the desired outcomes are given relative weights and listed on one axis. The measurable behaviors that affect the outcomes are listed on another axis. The degree to which the outcomes are affected by the behaviors can be assessed and then multiplied by the weight of the outcome to judge the relative importance of each behavior.

This type of decision matrix is applicable to curriculum development. When the best opportunities for improvement are listed as desired outcomes on one axis, and various Workshop activities are listed on a second axis as measurable student behaviors, we can assess the importance of student-run Workshops in developing the skills sought by industrial employers. It seems clear that student-run Workshops can add the preparation sought by industrial employers; eight of the top ten opportunities match directly with the goals and methods of the Workshop Project!

A pedagogy that advocates peer-led Workshops can benefit from collaboration among faculty and industrial technologists. Industrial technologists can assist faculty in developing Workshops far more easily than they could assist in traditional lecture and laboratory courses. Industry professionals can provide designs for technical modules and devices that focus on team efforts. A university-industry team approach that focuses on the top ten opportunities for improvement would be a unique approach to curriculum development. Of the twenty-five respondents, fifteen indicated that they are willing to contribute to the design of Workshops related to their area of expertise.

# *Appendix I*
# *Group Methods for Workshops: Specific Practice*

## Reflective Problem Solving

***Pair Problem Solving (Ronald Narode, Portland State University).*** Although there are many types of cooperative learning group activities, pair problem solving seems to maximize on-task behavior by students and affords the greatest opportunity for learning by *all* members of the group (Malter, Narode, and Davenport 1994). It may also be the easiest grouping to organize and manage. The model for the pair problem-solving method comes from Whimby and Lochhead (1981, 1986). These texts also provide excellent sources of problems for training students in the pair problem-solving technique.

The method of instruction using pair problem solving incorporates two key notions:

- constructivism - the idea that students must construct knowledge for themselves
- metacognition - the idea that the vehicle for construction of knowledge is self-reflection

Research indicates that the problems that students work on to develop conceptual understanding are often insufficient for learning. The various relationships among concepts and ideas, which constitute a conceptual web of understanding, are best developed and discerned by active discussion and debate with others (Von Glaserfeld 1988). Although knowledge is constructed individually, it is corroborated largely through consensus, and consensus building is a social activity.

The method assigns *solver* and *listener* roles to a pair of students. The *solver* is responsible for articulating all ideas as they occur. The *listener* tries to understand the process and the solution. Encouraging students to verbalize their thoughts requires them to examine their ideas. They must evaluate those ideas in light of another person's interpretation of what they are saying. Requests for clarification help students catch their errors or shore up good ideas that were still tentative. By exchanging roles of problem solver and listener, students have the opportunity to learn the related skills of explaining and listening.

In the pair problem-solving classroom, the teacher serves as a coach, moving from one pair to another, listening to their discussions and probing student solutions and conceptions with questions rather than answering questions. A student's answer is not acknowledged as either right or wrong. Instead, the instructor listens to the reasons for the answer and either agrees that the reasons make sense, asks for more elaboration, or asks more questions to help the student think about the problem in a different way. Often, the instructor asks the *listener* to explain the other student's solution and to explain why he or she does or does not agree with it.

*Instructions to the solver.* The approach requires that one student solve a problem by reading it aloud to the other student (the *listener)* and by verbalizing all thoughts on the problem as they occur. The *solver* does all the writing and all the talking about the problem. Meanwhile, the *listener* must suspend his or her own problem-solving activities and concentrate on understanding the *solver's* solution.

*Instructions to the listener.* The listener has a difficult task.

- *Listen carefully*: ask the *solver* to repeat statements or to slow down.
- *Encourage vocalization*: ask, "What are you thinking?" and "Can you explain what you are writing?"
- *Ask for clarification*: for example, "What do you mean," and "Can you say more about that?"
- *Check for accuracy*: ask, "Are you sure about that?" Several warnings are offered: do not give hints, do not solve the problem yourself, do not tell the *solver* how to correct an error.

*Instructions to the instructor.* The instructor can promote metacognitive activity by asking questions that require students to reflect on their thought processes. Four such strategies are to (Confrey 1985)

- ask students to discuss their interpretations of the problem;
- ask the students to describe precisely their methods of solution;
- ask students to defend their answers and their solutions; and
- ask students to retrace their steps to review the process they used to solve the problem.

The teacher should direct all questions to the *listener* and not to the *solver*. If the *listener* cannot explain the *solver's* solution, then the *listener* should be directed to ask the *solver* to repeat the solution. The teacher should probe for uncertainties or confusions and indicate agreement as appropriate. When the pair has come to some resolution, they should be encouraged to present their solution to the class. In this manner, the roles of *solver* and *listener* will be reinforced.

Students rarely see experts solve problems, much less hear them solve problems aloud. In addition to listening to students, teachers should model expert problem solving for their students. By describing their thought processes, instructors can demonstrate the process of thinking aloud and reveal the dead ends, mistakes, and corrections that characterize real problem solving.

The pair problem-solving process forces students to articulate their ideas to one another in a manner that causes students to explore their own understanding. In their attempts to explain their solutions and to defend against argument, students develop representations, analogies, limiting cases, multiple hypotheses, straw men, contradictory evidence, and the like. All of these activities require that the conceptual web of understanding grow in subtlety and complexity in the thought processes of the *solver* and the *listener*.

***Creating a flowchart (David K. Gosser, City College of New York).*** Once a problem has been solved it is very instructive to look back on the process that led to the solution. Beginning students are usually not self-conscious about their own mental processes. "I just did it" is a typical description. A retrospective analysis can put form and logic on the process. One way to do this is to ask the students to construct a flowchart of their steps to the solution. The chart makes the thinking process visible. This method would work well to wrap-up the stoichiometry problem (Chapter 3, p. 24) or Observation-Deduction problems (Chapter 3, p. 28). Constructing a flowchart to summarize a pair problem solving exercise would also work well. Subgroups of students could work together to solve problems and then create flowcharts that described their real problem-solving processes. The flowcharts could be exchanged, used and critiqued by other subgroups, or presented to the entire group for discussion.

The following flowchart illustrates an expert view of the process of problem solving. The actual process is more flexible than the flowchart indicates; for instance, it may be in the middle of the design process that one realizes the need for more information.

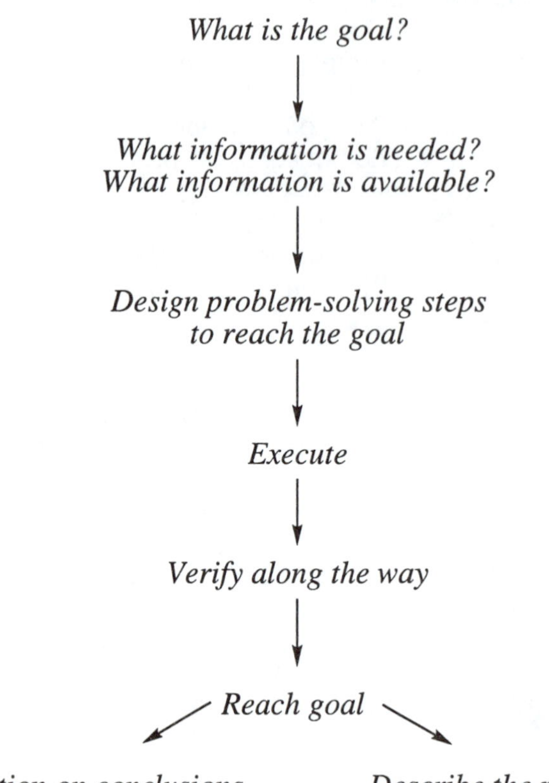


**Interactive Participation**

***Round-robin (David G. Gosser, City College of New York).*** The round-robin method for structuring group interaction and discussion has been described in other parts of this Guidebook. The Lewis-dot problem (Chapter 3, p. 00) provides a specific example. In addition to getting everyone involved, round robin focuses the group's attention on specific issues and rations the time among the participants, thereby preventing individuals from dominating the discussion.

Although the stepwise problems that are suitable for a round robin treatment are not the highest order of problem solving, they take on an added dimension in the hands of a skilled Workshop leader. The method is most powerful when the leader encourages reflection and discussion of the nuances of the "well-defined" steps. For example, *which electrons are available for bonding?* (incorrectly defined in some texts) or *why doesn't this particular example follow the octet rule?* It turns out that apparently simple algorithms lead to quite subtle features of problem-solving communication and understanding.

***Interview exam (Andrei Lalla, Workshop Leader, City College of New York).*** Students often need a mechanism for recovering from a misstep (e.g., a disappointing exam). The interview exam is a chance for students to make a good showing and obtain some extra credit. Students are informed that the interview exam is an oral presentation to the Workshop group. The faculty member gives the student the problems for the upcoming Workshop and a presentation guide (outlined below). The student prepares written answers to the Workshop problems beforehand. These answers are reviewed by the leader at the start of the Workshop. Once the written work is approved, the leader chooses a problem for the student to present to the group. After the oral presentation the leader asks the group for questions and authentication of the answers.

Students are graded on the following criteria:

- Written work: answers, calculations, and explanations as demonstrated in the written work
- Oral presentation: the student is required to follow these three steps:
    - i. read the question to the group,
    - ii. solve the problem on the blackboard, providing a step-by-step explanation of the solution, and
    - iii. answer questions posed by the group and the Workshop leader.

***Mendeleev game (Andrei Lalla, Workshop Leader, City College of New York).*** To help prepare for an exam, students in the Workshop are divided into two subgroups; each subgroup is a "Jeopardy" team. Five or six categories, dealing with major topics covered in lecture, are delineated as five questions and answers. Each question in a category is assigned a point value ranging from 100 to 500. The faculty member works with the Workshop leaders to create the appropriate categories, questions, and answers.

Before the session starts, the Workshop leader tapes the questions face down on the blackboard. The point value is written on the back of the question, facing the students. The questions are organized by categories with appropriate headings. The Workshop leader acts as scorekeeper, and the faculty member is the emcee.

Each subgroup chooses a team captain, and the two flip a coin to see who will choose the first question. *Both* subgroups now work on finding the answer to the same question and writing it down on paper. The first group to arrive at a solution to the question yells ***Mendeleev!***, and the Workshop leader places a check mark on the paper. This ensures that there is no possibility of changing answers during the game. At the same time, the second group should complete their written solution and yell ***Mendeleev!***

The first group to yell ***Mendeleev!*** then presents their answer and if correct, is awarded 200 points plus the point value of the question. However, if the answer is incorrect, 50 points are deducted. The second group answers the question only if the first group's answer was incorrect. If the second group answer correctly, they get 100 points and the point value of the question. This process is continued until all the categories are completed and one team wins outright (American Chemical Society 1997).

## References

American Chemical Society (1997). *Teaching Chemistry, 1997: Undergraduate Chemistry Curriculum Reform - Its Effect on High School and College Level Teaching.* Washington, D.C.: ACS Satellite TV Seminar, November 3, 1997.

Confrey, J. (1985). *A Constructivist View of Mathematics Instruction, Part II: An Empirical Examination.* Paper presented at the Annual Meeting of the American Educational Research Association, Chicago, Ill.

Malter, M.; R. Narode, and L. Davenport (1994). A Comparison of Pair and Small Group Problem Solving in Middle School Mathematics. *Proceedings of the North American Chapter of Psychologists of Mathematics Education Fifteenth Annual Conference*, San Jose State University, Vol. 2, pp. 26-32.

Von Glaserfeld, E. (1988). Cognition, Construction of Knowledge, and Teaching. *Synthese* (special issue on philosophy of science and education).

Whimbey, A. and J. Lochhead (1981). *Developing Mathematical Skills.* New York: McGraw-Hill.

Whimbey, A. and J. Lochhead (1986). *Problem Solving and Comprehension.* Hillsdale, N.J.: Lawrence Erlbaum Associates.

# *Appendix II*
# *Workshop Leader Training: Specific Practice*

**Vicki Roth, University of Rochester**

**Troubleshooting**

> *I appreciate just how important good leadership is for efficient group progress. When I look at the role I play in the Workshop, and look back at the experiences I have had there, I can see where my leadership or lack of leadership has had a profound effect.*
>
> *Kathy Castellano, University of Rochester leader*

The training of leaders and the implementation of the Workshops themselves is not an exact science; people being what they are, all sorts of wonderfully strange problems can emerge during the course of the semester. Some of these are so idiosyncratic that they just have to be dealt with on a case-by-case basis, but other issues are more predictable. Here are a few that we have seen over the course of our work with this model, along with some suggestions for addressing them.

***Problems within the Workshop sessions.*** *Unprepared students.* A common complaint of Workshop leaders (and professors everywhere) is that their students come to the sessions without having done the requisite preparation. Leaders are frustrated because they feel forced to recapitulate what was said in the text or what was to have been learned in homework problems.

Leaders need to be encouraged to resist the tug toward lecturing the group. Certainly, there are times when they will want to give microlectures about specific concepts; a quick review of $S_N1$/E1 and $S_N2$/E2 or the second law of thermodynamics may be just the right thing at certain points. But unless the participants see and feel the impact of their lack of preparation, that is, that they cannot address the Workshop problems well during the sessions, they have little natural incentive for change.

But we should not assume that students know how to change, just because they decide they want to. Learning center specialists can testify that students often do not have the strategies needed to approach their studies well. Students open their books and pull out their problem sets, give things a try, get frustrated by the complexity of the material, and cave in all too quickly, so sometimes leaders need to give a few study skills pointers. At other times, students may best be served by a referral to the study skills center.

*A sluggish group.* Groups, just like people, tend to have personalities of their own, and sometimes they have fairly lethargic ones. If a group fails to catch fire after the first get-acquainted sessions have passed, the leader needs to take action. Frequently a change in seating arrangements or even switching rooms is called for; people who will not talk much in one format may open up in another. Or maybe this group is hungry; Workshops that meet during a typical meal time should be encouraged to brown bag it or pool funds for pizza. Perhaps there is one member of the group who tends to shut the other students down; leaders may need to find strategies to rein in this member so the others can find a comfortable place in the group. It is always fair game to ask the participants directly what will help them be more productive in these sessions. A conversation of this sort may be a bit uncomfortable at first, but honest communication can lead to a substantial payoff for everyone.

If these first-order ideas fail to galvanize the group, it is time for an observer to spend some time in the sessions; often a fresh pair of eyes can see some room for growth.

***Problems of individual Workshop members.*** *Absentees.* As good as we believe the Workshop model to be, it cannot help students if they are not showing up. We keep careful track of attendance in our sessions, not as a way to enforce rewards and punishments but rather as a means of monitoring the health of the groups. We encourage leaders to respond promptly when a student has missed a session; a friendly "Hi, what's up?" call or an E-mail message is a good way to open a line of communication with a student who may not be fully a part of the group. We also encourage leaders to avoid looking at these absences too narrowly ("That student is too lazy" or "I must not be a very good leader if that student doesn't want to come to my group"). Absenteeism stems from many causes; sometimes students *are* just lazy, but in the other extreme, they may have missed a session because they were taking family members to the hospital or dealing with other serious life issues. We tell our leaders that everyone in the world deserves at least one rescue call, regardless. Sometimes, it is just this moment of personal attention from the leader that helps reconnect a troubled or disengaged student.

If absenteeism is an issue for the group as a whole, there is likely to be a system problem in play. This is the time for the program directors or the student coordinators of the program to make an offer to attend a session or two and directly help the leader get the group back on track.

*At times, one student tries to really confound me. Whether this is done intentionally I do not know.*
*George Buse, University of the Pacific leader*

*Hostile students.* We are putting our leaders in the educational front line, so they may from time to time encounter a student whose bad attitude is more than mere crankiness. When this happens, your leader needs some support, first through discussion with you and with the other leaders. It helps to talk about why students may be acting belligerently (e.g., fear of looking foolish in front of other students, family troubles, emotional problems). Understanding possible causes helps leaders feel a little less put on the spot and it allows them to avoid a purely defensive stance and instead think through possible responses instead. Sometimes, however, the hostility becomes more than what peer leaders should be expected to handle independently. In these cases, they need to know that we as program directors will support them. In these cases the best way we can demonstrate this is by intervening ourselves, perhaps by visiting the group sessions or speaking directly to the student in question.

*Personal problems.* We would like to erect a protective bubble over our students to allow them to concentrate on the development of their academic selves, but it is a rare student who makes it all the way through the college years without the intrusion of some outside issues. We know that many students are coping with illness, financial distress, divorce, child-care crises, problems on the job, and so on. It is common for troubled students to unburden themselves to our leaders, not surprisingly, since these peer leaders are usually concerned and caring people. In turn, often the temptation for the leaders is to take on too much of the responsibility for helping students work their way through these sorts of problems. We need to help them understand once again the nature of boundary setting. We do not want our Workshop leaders to think of themselves as therapists; not only could this role overwhelm the leaders but it could lead to all sorts of misdirection. Instead, they should be familiar with ways to identify the problem at hand, to know the location of campus resources, and to master effective referral strategies. This preparation merits a spot on the training schedule.

***Problems with leaders.*** Despite our best recruiting and interviewing strategies, we occasionally hire someone who does not turn out to be a good fit for the job. The most common problem is that we have someone who continues to function like a recitation leader; that is, by holding forth in front of the class at length, or by running the Workshop like a question-and-answer session. Probably the most powerful way of addressing this issue is to have the leader sit in on a few Workshop sessions in which the model is being implemented more effectively. Mere talk about the idea of the Workshop can pass some new leaders by, but change can happen when people are able to witness what this model can do.

At other times it may be useful to have an observer, perhaps one of your more experienced leaders, visit the sessions about which you are concerned and provide feedback to the leader. Some training programs build in the expectation that all leaders will take turns observing one another's sessions and providing feedback. This method helps to avoid a sense of "policing" troubled groups; if everyone is being observed, no one is being singled out.

**Some Guaranteed Icebreakers**

For groups of twenty-five or more, you might like to play "Introduction Bingo." Prepare a sheet with questions that match the personal characteristics of your leaders, for example, "This person is from a non-English-speaking country" or "This person worked in a genetics lab this summer." Even better are those funny, but harmless, characteristics that you might know about some of your leaders, like "This person REALLY hates spiders."

After handing out the bingo sheets to your students, inform them that their job is to find someone who fits the profile of each question. Then turn them loose. In a few minutes, you will have a very noisy roomful of people getting to know one another. Once they have located a person who fits one of the descriptive statements, they introduce themselves and have their "match" sign the appropriate space on the bingo sheet. The first person who obtain all the answers either across, down, or diagonally calls "bingo" and is the winner. Provide some sort of reward to the winner - the sillier the better.

Here are a couple of additional warm-up exercises from *The Encyclopedia of Icebreakers* by S. Forbess-Green:

*Three Truths:* Have the members of the group share four things about themselves, one of which is false. Those who can pick out the whoppers win a piece of candy.

*Name Game:* Have everyone sit in a circle. The first person states his/her name and two personal descriptions. The next person then introduces the first person and states these two things again. The game proceeds around the circle, with each person introducing all who came before and their two descriptors.

**Samples**

Examples of specific invitations, assignments, questionnaires, and agenda that we have found useful are given on the following pages.

## *Sample Interest Meeting Invitation*

Dear <<first name>>,

Your work and accomplishment in your chemistry course this past semester were simply wonderful; please accept my very best congratulations. My experience tells me that your accomplishment was the result of your engagement with the course and the subject. In turn, that engagement must have come from your belief that the material was interesting and worth learning. I write now to invite you to help me try to get next year's class hooked on this course.

In particular, I think that the Workshop is a very successful format for actively engaging students with the subject material. As you heard me say many times, passive observers do not learn to think about chemistry. Next semester, I want to continue to provide Workshop opportunities for all the students in the course. Since the best size for the groups seems to be about eight students, I will need about (#) leaders. Students like you, with clear records of engagement in the course, are my best candidates for leadership.

Your primary responsibility as a Workshop leader will be to guide your students to the active engagement with the material that worked so well for you. Although you will have to be able to deal with questions about the problems that are given in the Workshop, you will not be expected to be an "authority" on chemistry. Rather, you will be an experienced student who did exceedingly well in the course and is now trying to help others do exceedingly well also. Each Workshop session will last two hours. I will provide a set of problems for each session.

As part of this program, you will need to enroll in a two-credit course offered by the director of the learning center to teach you how to be a Workshop leader. As part of the course, I will also review the ideas and methods behind the problems.

I think that there are several kinds of potential rewards for you. First, I think that you will find that your identification with our school and with your own education will change in significant ways as you take an active role in helping others learn. Second, I think that you will be surprised by the deepening of your own understanding of chemistry that will come from your involvement as a Workshop leader. Third, I think you will find great personal satisfaction from guiding other to understanding. And, finally, the job pays $xxx for the semester!

I hope that you will think about this opportunity in the next few weeks. The director of the learning center and I will hold an interest meeting on (date, time, location) to give you a chance to ask questions. The meeting is well before registration so that you can think about being a Workshop leader as you plan for your next semester.

Sincerely,

## *Sample Workshop Leader Job Description*

*(distributed at interest meeting)*

A Workshop leader:

1. Conducts weekly two-hour Workshop sessions as scheduled.
2. Attends the weekly Workshop training course and completes all assigned work for the course.
3. Prepares for Workshop sessions.
4. Informally evaluates the progress of individual group members via leader logs.
5. Maintains attendance records for the Workshop.
6. Participates in debriefing surveys and discussions about the Workshops via leader logs.

Workshop leaders will be paid $xxx for the semester. The Workshop responsibilities begin on (starting date) and continue until (ending date).

## *Sample Workshop Leader Application*

Date:

Name: Social Security #:

Campus address: Phone:

Permanent address: E-mail address:

Where can you be reached during the summer?

Your status next semester: Fr So Jr Sr Grad Other

Academic major: GPA:

*******************************************

Please answer the following questions on a separate page:

1. Why do you want to be a Workshop leader for this course?
2. Previous tutoring experience? Workshop experience? Related experience?
3. What skills qualify you to be Workshop leader?
4. How many hours per week would you be willing to devote to the Workshop program during the academic year?
5. Describe any work experiences you have had on campus.

Signature:

## *Sample Workshop Leader Contract*

*(distributed after final candidates are selected)*

A Workshop leader

1. Conducts weekly two-hour Workshop sessions as scheduled.
2. Attends weekly the Workshop training course and completes all assigned work for the course.
3. Prepares for Workshop sessions.
4. Informally evaluates the progress of individual group members via leader logs.
5. Maintains attendance records for the Workshop.
6. Participates in debriefing surveys and discussions about the Workshops via leader logs.

Workshop leaders will be paid $xxx for the semester. The Workshop responsibilities begin on (starting date) and continue until (ending date).

I can meet the responsibilities of a Workshop leader for (course) in (semester).

Name (print): SS#:

Name (sign): Date:

Faculty: Date:

## *Sample Agenda for a Workshop Leader Retreat*

**Day One**

| | |
|---|---|
| 9:00- 9:30 am | Breakfast |
| 9:30 - 10:00 am | Introduction to Workshops |
| 10:00 - 10:30 am | The role of the leader |
| 10:30 - 11:30 am | What is your learning style? |
| 11:30 - 12:00 pm | Icebreaker |
| 12:00 - 1:00 pm | Lunch |
| 1:00 - 2:00 pm | Problem-solving techniques |
| 2:00 - 3:00 pm | What are your concerns? |

**Day Two**

| | |
|---|---|
| 9:00 - 9:30 am | Breakfast |
| 9:30 – 9:45 am | Workshop leaders as future teachers |
| 9:45 - 11:00 am | Addressing Workshop leaders' concerns |
| 11:00 - 12:00 pm | Applying problem-solving techniques in Workshops. (round robin and pair problem solving) |
| 12:00 - 1:00 pm | Lunch |
| 1:00 - 2:00 pm | The Internet - Workshop web site and e-mail addresses |
| 2:00 - 2:45 pm | Journals - reflective writing |
| 2:45 - 3:00 pm | Administrative issue and student stipends |

## *Sample Syllabus for a Two-Credit Leader Training Course*

***Welcome to Issues in Group Leadership***. This course is designed to provide training and support to Workshop leaders. Our discussion in this course will include basic review of the Workshop modules and practical and theoretical aspects of Workshop leadership. It will also offer all of us the opportunity to explore individual topics of interest.

***Requirements and Grading:***

| | |
|---|---|
| Class attendance and participation | 30% |
| Short assignments | 20% |
| Weekly log | 20% |
| Individual project | 30% |

***Reading Materials:*** Most of our reading for this course will be distributed in class. Occasionally we may need to put longer materials on reserve in the library.

***Individual Projects:*** The individual project will allow you to explore a topic of special interest to you and will provide all of us the benefit of one another's experience. In addition, our project will give you the opportunity to practice making presentations in the format you will most likely be encountering in your first professional conferences.

These individual projects will include *poster sessions* and a matching 5-8 page *written report* (more information to follow).

***Course Schedule:***

| | | |
|---|---|---|
| Session 1 | Getting acquainted<br>Reviewing the syllabus<br>Planning for the first Workshop<br>Understanding the attendance system | Understanding the role of the leader<br>Confirming our schedules |
| Session 2 | Reviewing the first meeting<br>Assignment: Observe student behavior in large lectures | Group dynamics |
| Session 3 | General debriefing<br>Project assigned<br>Assignment: Write a homework problem that taps algorithmic thinking and another that demands conceptual understanding | Developing a statement about the ethics of group leadership |
| Session 4 | General debriefing<br>Assignment: Observe a peer's Workshop session and provide feedback | Learning theory and the Workshop |

| Session | Activities | Topic |
|---|---|---|
| Session 5 | General debriefing<br>Library support for projects<br>Assignment: Teaching a concept to different types of learners | Differences in learning styles<br>Making a good referral |
| Session 6 | General debriefing<br>Handling midterm evaluations | Science pedagogy and the Workshop |
| Session 7 | General debriefing<br>Constructing a good poster | Student development and the Workshop |
| Session 8 | General debriefing | Race, class, and gender and the Workshop |
| Session 9 | General debriefing | Using concept maps and graphic aids |
| Session 10 | General debriefing<br>How to write a good paper | Motivation and the Workshop |
| Session 11 | Poster sessions (or project presentations) | |
| Session 12 | Poster sessions (or project presentations) | |
| Session 13 | Papers due<br>Conducting end-of-the-year evaluations | Wrapping up the semester |

## *An Observation Assignment: Helping One Another*

**Once you have completed this form, please *make a copy of it. Give the original to your classmate, and turn in the copy next week in class.***

***Here's the plan:***

1. Select a Workshop to observe next week.
2. Inform the group leader of your intention to visit that session.
3. When you arrive at the session, introduce yourself casually. Try something like, "Hi, I'm a leader for another group. I wanted to sit in to compare notes."
4. Let the discussion proceed as much as possible as though you weren't there. Be a fly on the wall. Right after the group is over, write down your impressions so you don't lose the immediacy of what you want to say.

**Your name:** **Your classmate's name:**

**Date you observed the group:** **Location of the group:**

1. What did you think about the participants' seating arrangement during this session?
2. What happened to help everyone get down to work?
3. What did you observe about group interaction? How much did individual students pay attention to one another?
4. What sorts of things happened when someone's understanding broke down?
5. How well did the students seem to understand the concepts included in this module by the end of the session? What gave you this impression?
6. What suggestions do you have for the leader for upcoming sessions?
7. What was the very best thing about this Workshop?
8. Anything else?

## *Self-Rating Checklist*

*Please rate your leadership and your Workshop sessions according to the following questions. Feel free to make additional comments and observations at any point. Note that you are* NOT *being asked to put your name on this form, so please be candid in your replies.*

| | | Very Seldom | Sometimes | Often | Almost Always |
|---|---|---|---|---|---|
| 1. | I understand the goals of the Workshop program. | 1 | 2 | 3 | 4 |
| 2. | I understand the professor's goals for the modules and for the course as a whole. | 1 | 2 | 3 | 4 |
| 3. | I prepare well for the Workshops. | 1 | 2 | 3 | 4 |
| 4. | I can get the Workshop sessions started easily. | 1 | 2 | 3 | 4 |
| 5. | I am happy with the amount of control I have in the Workshop (not too much, not too little). | 1 | 2 | 3 | 4 |
| 6. | I am patient. | 1 | 2 | 3 | 4 |
| 7. | All the students participate on a regular basis. | 1 | 2 | 3 | 4 |
| 8. | I am adept at keeping the conversation going among the students. | 1 | 2 | 3 | 4 |
| 9. | I am good at asking questions that help students approach the problems. | 1 | 2 | 3 | 4 |
| 10. | My students look to me for help the appropriate amount of time (not too often, not too seldom). | 1 | 2 | 3 | 4 |
| 11. | The Workshop students talk easily with one another. | 1 | 2 | 3 | 4 |
| 12. | I am able to tell people they are incorrect in a constructive way. | 1 | 2 | 3 | 4 |
| 13. | I am confident about explaining the material when it is appropriate for me to do so. | 1 | 2 | 3 | 4 |
| 14. | I know how to handle it when someone asks me a question I can't answer. | 1 | 2 | 3 | 4 |
| 15. | My Workshop stays on task the right amount of time each week. | 1 | 2 | 3 | 4 |
| 16. | I can get the students back on track when they get distracted. | 1 | 2 | 3 | 4 |

| | | | | | |
|---|---|---|---|---|---|
| 17. | I am able to break the tension when needed; I can keep the stress within the group at a manageable level. | 1 | 2 | 3 | 4 |
| 18. | I help students see connections between old and new material. | 1 | 2 | 3 | 4 |
| 19. | I take different learning styles into account when I plan a Workshop session. | 1 | 2 | 3 | 4 |
| 20. | I treat all students fairly. | 1 | 2 | 3 | 4 |
| 21. | I keep tabs on individual students' progress with the Workshop problems. | 1 | 2 | 3 | 4 |
| 22. | I know how to make referrals to other campus resources when necessary. | 1 | 2 | 3 | 4 |
| 23. | I think about the possible impact of race, class, and gender issues on students' learning. | 1 | 2 | 3 | 4 |
| 24. | If I had a student with a disability in my Workshop, I would know how to plan for support for him/her in my Workshop sessions. | 1 | 2 | 3 | 4 |
| 25. | I maintain a positive attitude. | 1 | 2 | 3 | 4 |

**Other comments:**

# *Appendix III*

## *Evaluation: Specific Practice*

**Leo Gafney**

**The following questionnaire is designed to be SCANTRON compatible.**

## *STUDENT SURVEY*

**Peer-Led Team Learning** **Institution** ______________________________

**Course name** ___________________________ **Professor** ______________________________

**For each item, circle the number that corresponds to your response: 1 = strongly disagree; 2 = disagree; 3 = neutral (no opinion); 4 = agree; 5 = strongly agree.**

| |  | | | | | |
|---|---|---|---|---|---|---|
| 1. | The Workshops are closely related to the material taught in the lectures. | 1 | 2 | 3 | 4 | 5 |
| 2. | Workshops help me do better in tests. | 1 | 2 | 3 | 4 | 5 |
| 3. | Interacting with the Workshop leader increases my understanding. | 1 | 2 | 3 | 4 | 5 |
| 4. | The Workshop materials are helpful preparation for exams. | 1 | 2 | 3 | 4 | 5 |
| 5. | The Workshop materials are more challenging than most textbook problems. | 1 | 2 | 3 | 4 | 5 |
| 6. | I believe that the Workshops are improving my grade. | 1 | 2 | 3 | 4 | 5 |
| 7. | I regularly explain problems to other students in the Workshops. | 1 | 2 | 3 | 4 | 5 |
| 8. | Interacting with the other group members increases my understanding. | 1 | 2 | 3 | 4 | 5 |
| 9. | I would recommend Workshop courses to other students. | 1 | 2 | 3 | 4 | 5 |
| 10. | In the Workshops I am comfortable asking questions when I do not understand something. | 1 | 2 | 3 | 4 | 5 |
| 11. | The lecturer encourages us to participate in the Workshops. | 1 | 2 | 3 | 4 | 5 |
| 12. | The Workshops are often dominated by one or two students. | 1 | 2 | 3 | 4 | 5 |
| 13. | Noise or other distractions make it difficult to benefit from the Workshops. | 1 | 2 | 3 | 4 | 5 |
| 14. | Students who are uninterested or unmotivated make it difficult for others to benefit from the Workshops. | 1 | 2 | 3 | 4 | 5 |
| 15. | I feel comfortable with the Workshop leader. | 1 | 2 | 3 | 4 | 5 |
| 16. | The Workshop leader is well prepared. | 1 | 2 | 3 | 4 | 5 |
| 17. | I am uncomfortable asking questions in the lecture. | 1 | 2 | 3 | 4 | 5 |
| 18. | The Workshops are a big help in solving problems. | 1 | 2 | 3 | 4 | 5 |
| 19. | I would like to be a Workshop leader in the future. | 1 | 2 | 3 | 4 | 5 |
| 20. | In the Workshops I enjoy interacting with the other students. | 1 | 2 | 3 | 4 | 5 |
| 21. | The Workshop experience led me to join formal or informal study groups related to other courses. | 1 | 2 | 3 | 4 | 5 |

**Please turn over and complete the items on the other side.**

**Circle the appropriate response on the 1-5 scale, as specified in the following questions.**

22. On average, I spend the following number of hours per week studying (in addition to time spent at lectures and Workshops):

    1. 0–2 hours   2. 2–4 hours   3. 4–6 hours   4. 6–8 hours   5. 8–10 hours

**This next item is about the materials used in the Workshops.**
**Use the following scale: 1 = materials do not meet this objective at all; 2 = materials somewhat meet the objective; 3 = materials meet the objective rather well; 4 = materials meet this objective very well; 5 = materials are excellent for meeting this objective:**

The materials are:

| | | | | | | |
|---|---|---|---|---|---|---|
| 23. | well connected with the lecture | 1 | 2 | 3 | 4 | 5 |
| 24. | challenging | 1 | 2 | 3 | 4 | 5 |
| 25. | developed to review fundamentals | 1 | 2 | 3 | 4 | 5 |
| 26. | useful for group work | 1 | 2 | 3 | 4 | 5 |
| 27. | motivational | 1 | 2 | 3 | 4 | 5 |
| 28. | helpful for individual study | 1 | 2 | 3 | 4 | 5 |
| 29. | useful for reinforcing concepts | 1 | 2 | 3 | 4 | 5 |

**Rate each of the following activities according to amount of Workshop time devoted to the specified activity.**
**Use the following scale: 1 = almost no time; 2 = a small amount of time; 3 = a moderate amount of time; 4 = a large amount of time; 5 = most of the time.**

| | | | | | | |
|---|---|---|---|---|---|---|
| 30. | The Workshop leader presents ideas and methods. | 1 | 2 | 3 | 4 | 5 |
| 31. | The leader responds to student questions. | 1 | 2 | 3 | 4 | 5 |
| 32. | Students work on problems in pairs or small groups. | 1 | 2 | 3 | 4 | 5 |
| 33. | Students work on problems alone. | 1 | 2 | 3 | 4 | 5 |
| 34. | Students present solutions. | 1 | 2 | 3 | 4 | 5 |
| 35. | Hands-on activities. | 1 | 2 | 3 | 4 | 5 |
| 36. | Technology and computer simulations. | 1 | 2 | 3 | 4 | 5 |

**Thank you for your participation.**

## *LEADER SURVEY*

**Peer-Led Team Learning** **Institution** ______________________

**Course name** ______________________ **Professor** ______________________

1. How often do Workshops meet? __________
2. What is the scheduled length of a Workshop meeting? __________
3. On average, how long do you usually meet? __________
4. How many students are enrolled in your Workshop? __________
5. On average, how many students usually attend a Workshop? __________
6. What do you think is the best number of students for a Workshop? ______________
7. Attendance at the Workshop ( is, is not ) a course requirement?
8. About how much of your time per week is taken by Workshop preparation and activities, not including the Workshop itself? ____________
9. Please describe the activities as they take place in a typical Workshop?

**Please rate each of the following activities according to Workshop time devoted to it. Use the following scale: 1 = almost no time; 2 = a small amount of time; 3 = a moderate amount of time; 4 = a large amount of time; 5 = most of the time.**

| | | | | | |
|---|---|---|---|---|---|
| 10. The Workshop leader presents ideas and methods. | 1 | 2 | 3 | 4 | 5 |
| 11. The leader responds to student questions. | 1 | 2 | 3 | 4 | 5 |
| 12. Students work on problems in pairs or small groups. | 1 | 2 | 3 | 4 | 5 |
| 13. Students work on problems alone. | 1 | 2 | 3 | 4 | 5 |
| 14. Students present solutions. | 1 | 2 | 3 | 4 | 5 |
| 15. Hands-on activities such as use of models. | 1 | 2 | 3 | 4 | 5 |
| 16. Use of technology or computer simulations. | 1 | 2 | 3 | 4 | 5 |

17. Are Workshop problems good preparation for tests? Please describe.

18. Do Workshop materials include challenging problems? Please describe.

**Please turn over and complete the items on the other side of this page.**

19. Were the Workshop materials too difficult or too easy for students in your group? If so, what did you do?

**This item is about the materials used in the Workshops.**
**Use the scale from 1 to 5 according to how well they meet each objective: 1 = materials do not meet this objective at all; 2 = somewhat meet the objective; 3 = meet the objective rather well; 4 = meet this objective very well; 5 = materials are excellent meeting this objective**:

The materials are:

| | | | | | |
|---|---|---|---|---|---|
| 20. well connected to the lecture | 1 | 2 | 3 | 4 | 5 |
| 21. challenging | 1 | 2 | 3 | 4 | 5 |
| 22. developed to review fundamentals | 1 | 2 | 3 | 4 | 5 |
| 23. useful for group work | 1 | 2 | 3 | 4 | 5 |
| 24. motivational | 1 | 2 | 3 | 4 | 5 |
| 25. helpful for individual study | 1 | 2 | 3 | 4 | 5 |
| 26. useful for reinforcing concepts | 1 | 2 | 3 | 4 | 5 |

27. What methods are used to get students working together?

28. What do you do for students having difficulty?

29. Did students sometimes discuss personal problems with you? If so, how did you respond to them?

30. What training and support are provided to leaders in how to run Workshops, for example in group dynamics or instructional processes?

31. What training and support are provided to the Workshop leaders in the knowledge of the discipline?

32. What training and support are provided to the Workshop Leaders in theories of learning and related methods of teaching?

33. What parts of student leader training have been most useful? What do you need more of?

34. How do you interact with the professor teaching the Workshop course?